Electromagnetic Geophysical Fields

Oleg Novik • Feodor Smirnov • Maxim Volgin

Electromagnetic Geophysical Fields

Precursors to Earthquakes and Tsunamis; Impacts on the Brain and Heart

Springer

Oleg Novik
IZMIRAN of the Russian Academy of Sciences
Moscow, The Moscow Area, Russia

Feodor Smirnov
IZMIRAN of the Russian Academy of Sciences
Moscow, The Moscow Area, Russia

Maxim Volgin
IZMIRAN of the Russian Academy of Sciences
Moscow, The Moscow Area, Russia

ISBN 978-3-319-98460-5 ISBN 978-3-319-98461-2 (eBook)
https://doi.org/10.1007/978-3-319-98461-2

Library of Congress Control Number: 2018960998

This Springer imprint is published by the registered company Springer Nature Switzerland AG
The registered company address is: Gewerbestrasse 11, 6330 Cham, Switzerland

The authors dedicate this book to
B. G. Novik (who died at the front in World War II in 1944),
L. I. Babenko (1930–1990),
P. S. Volgin,
and I. B. Mikhaylovskaya (1939–2007).

Preface

In the twenty-first century, hundreds of thousands of people have been killed by the tremendous energy released from deformations of the oceanic or continental lithosphere, i.e., earthquakes (EQs), whose seismic source originated beneath the ocean or land.

This has always been the case, but now the vulnerability of technology to these seismic processes is critically dangerous. The seismic disruption of the Fukushima I nuclear power plant resulted in radioactive pollution of large areas of sea and land, with deaths and injuries in the populace.

It is a sign that nuclear power plants are hardly able to replace the limited or depleted traditional sources of energy (gas, oil, hydroelectric energy, geothermal resources, etc.) without effective geological, geophysical, and geochemical prognostic science and, as a result, the stable prediction of land and marine EQs needed to shut down a nuclear plant before such a geodynamic event occurs.

The dynamics of the exterior geospheres—namely the solar-induced perturbations of the magnetosphere and the ionosphere, and the resulting electromagnetic (EM) fields (geomagnetic storms)—may disrupt the operation of power networks, air and ground transportation, and navigation, and may accelerate deterioration of pipelines, etc. The electrical system of human brains is also exposed to these fields, including the brains of those humans who control sophisticated technologies with severe consequences for any control error.

In this book, we consider theoretical questions of field interaction physics, including EQ precursors, and we also present the results of measurements of the influence of specific interactions, namely geomagnetic storms, on the EM processes of the human brain and heart. Mathematical problems of the theory of interaction of geophysical fields of different physical natures in land and marine EQ preparation processes (*precursory physics*) may not be considered in the frame of the classical initial boundary value problem formulation. Indeed, all geophysical fields are disturbed during the preparation processes because of the release of large portions of the energy of deformation and the impossibility of a purely mechanical process in the nonideal medium of a seismically active lithosphere zone. But physically different fields are described by differential operators of different mathematical types. The resulting field interaction mathematical model [i.e., the initial boundary value problem of the system of partial differential equations (PDEs)] may not be of a well-investigated classical type, though the authors apply a physically correct approach on the basis of the main physical principles and geological and geophysical data on the medium where the interaction is located. Therefore, investigations are needed to determine whether the interaction model is correct from the mathematical viewpoint, i.e., does it have a unique solution with continuous (by the proper functional norms) dependence on the input data of the physical problem under consideration? Usually, the input data are known to have some errors accompanying the measurements providing the data, and continuous dependence of the solution decreases the influence of this error.

In Chaps. 9, 10, and 11, the authors describe the simplest but most important parts of their mathematical investigations of the field interaction models. So, the practical recommendations in this book are based on physically and mathematically correct models, with the numerical characteristics of the computed signals being consistent with measurements published by other authors.

Let us note that the physical sense of our geophysical results related to the precursory physics, along with the computer graphics clearly illustrating this sense—namely the evolution of interacting fields of different natures—may be understood without mathematics related to the nonclassical initial boundary value problems. These problems arising in the precursory physics were formulated and investigated mathematically and then solved numerically by the authors to obtain geophysical results with mathematical justification. The authors believe that, among other scientific and educational reasons (especially for young researchers), this justification is needed because of the recommendations for a prognosis on the basis of the results obtained. Holistically, we are trying to correctly consider the dangerous interaction of geophysical fields, not only between themselves, but with humans as well, i.e., we are trying to discuss the elements of geo-biophysics with regard to human safety related to natural dynamical processes in the geospheres.

Moscow, Russia

Oleg Novik
Feodor Smirnov
Maxim Volgin

Acknowledgements

We are grateful to our parents and our excellent school teachers for our initial learning.

We are grateful to our excellent teachers from the Petersburg and Moscow State Universities.

We are grateful to our colleagues from the Moscow State Geology Prospecting Institute, the Moscow Medical Education Institute, and the Pushkov Institute of Terrestrial Magnetism, Ionosphere and Radio Wave Propagation (IZMIRAN), Russian Academy of Sciences, including the board of directors. We are grateful to the colaborators of Springer Nature for their work.

We are grateful for discussions, questions, and interest in our work.

IZMIRAN, Moscow, Russia
August 17, 2018

Oleg Novik
Feodor Smirnov
Maxim Volgin

Contents

Abbreviations

1D	One-dimensional
2D	Two-dimensional
3D	Three-dimensional
ACh	Acetylcholine
AM	Amplitude modulation
ANS	Autonomous nervous system
AP	Action potential
ASTER	Advanced spaceborne thermal emission and reflection radiometer
ATP	Adenosine triphosphate
AV	Atrioventricular
cAMP	Cyclic adenosine monophosphate
CF	Coherence function
CFV	Coherence function value
CIR	Corotating interaction region
CME	Coronal mass ejection
CNS	Central nervous system
CSEM	Controlled source of an electromagnetic field
CT	Computed tomography
CTD	Conductivity, temperature, depth
DART	Deep-Ocean Assessment and Reporting of Tsunamis
DC	Direct current
DMTS	Deep magneto-telluric soundings
Dst	Disturbance storm time
ECG	Electrocardiography
EEG	Electroencephalography
EM	Electromagnetic
EMF	Electromagnetic field
EMG	Electromyogram
EPSP	Excitatory postsynaptic potential
EQ	Earthquake

ERS	Event-related synchronization
FAC	Field-aligned current
FFT	Fast Fourier transform
FLC	Fraction of low coherence
fMRI	Functional magnetic resonance imaging
GPS	Global Positioning System
HSS	High-speed solar
IAGA	International Association of Geomagnetism and Aeronomy
IPSP	Inhibitory postsynaptic potential
IR	Infrared
IT	Information technology
IZMIRAN	Pushkov Institute of Terrestrial Magnetism, Ionosphere and Radio Wave Propagation
K_p	Planetary geomagnetic index
LRS	Low-resistivity structures
MEG	Magnetic encephalography
MHD	Magneto-hydrodynamic
MLPEEA	Magnetic location of a possible earthquake epicenter area
MMT	Marine magneto-telluric
MRI	Magnetic resonance imaging
MS	Magnetic storm
MTE	Magneto-thermo-elasticity
MTS	Magneto-telluric soundings
N	Nodal
NASA	National Aeronautic and Space Administration
NH	Nodal–His
NOAA	National Oceanic and Atmospheric Administration
NTS	Nucleus tractus solitarius
ODE	Ordinary differential equation
PDE	Partial differential equation
PET	Positron emission tomography
PNS	Peripheral nervous system
PRT	Proofreading test
rw	Rolling wave
SA	Sinoatrial
S.-D.	String–diffusion
SAR	Synthetic aperture radar
SDSE	String–diffusion system of equations
SEMTI	Seismo-electromagnetic-thermal interaction
SHEMTI	Seismo-hydro-electromagnetic-thermal interaction
SWPC	Space Weather Prediction Center
T	Temperature
Th	Threshold

TSUNPREDET	Multidisciplinary vertically distributed seaquake/tsunami precursors detection system
TV	Thermo-viscoelastic
UCLA	University of California, Los Angeles
ULF	Ultra-low-frequency
UT	Universal time
VHF	Very-high-frequency

Chapter 1
Low-Resistivity Structures in the Seismically Active Lithosphere Zones and Observations of Electromagnetic Signals of Seismic Origin

Abstract In this chapter, we cite the published results of the field investigations of other authors regarding the characteristics of (a) low-resistivity (i.e., low-ohmic) structures in seismically active lithosphere zones and (b) electromagnetic (EM) signals of seismic origin (seismo-EM signals). The typicality of similar structures in seismically active zones, according to numerous expeditions and the geological theory of ore genesis, is discussed as well.

These data may be considered as the geological and geophysical basis of the theory of seismo-EM interaction and earthquake (EQ) precursory signals (also referred to as precursors), which are discussed further in the subsequent chapters. Precursors are understood here as the measurable signals before an earthquake, i.e., without any estimations (these have been doubtful, recently) of their prognostic value.

Keywords Earthquake · Ore genesis · Low-resistivity structure · Precursor · EM signals

1.1 Introduction

The mechanism of generation of electromagnetic (EM) signals before an earthquake (EQ) is not well understood. Moreover, we believe there are different mechanisms of similar generation due to the large variety of seismically active geological structures and mechanical processes of their deformations resulting in an EQ. Nevertheless, not all of them are well known. Now it is necessary to determine the expression "to know well the mechanism of generation of EM signals before an EQ." Surely, it is possible to speak about the generation mechanism only for the given seismic lithosphere zone, because of the variety and differences. First of all, a reasonable mechanism must be based on the main particularities of the geological structure of a zone under consideration and the measured characteristics of the process of its deformations. The latter may not be purely mechanical (e.g., because of nonideal rheology of the lithospheric rocks) and is accompanied by generation of

O. Novik et al., *Electromagnetic Geophysical Fields*,
https://doi.org/10.1007/978-3-319-98461-2_1

physical fields of a different nature—namely, the following phenomena have been observed before an EQ: seismic (deformation), EM, and acoustic waves; changes in the temperature (T) and heat flow fields; hydrodynamic disturbances (nonstationary filtration and a flow of gases, e.g., radon); gravitational disturbances; changes in the fields of concentrations of the chemicals in the underground fluids and in ones of a borehole; etc.

It is possible to propose that we "know well the mechanism of generation of EM signals before an EQ" for the given seismically active region in the case when we know the geological and geophysical nature (or basis) of the mechanism, i.e., we are able to connect the observed disturbances of the geophysical fields with the particularities of the geological structure of the region and its seismic excitation. Then the proper domain of physics must be chosen to take account of these data and to formulate a mathematical model of the precursory signals generation according to the geological–geophysical basis and main physical principles. Then the mathematical investigation of the correctness of the model and its discretization for computations (i.e., algorithm, because an analytical approach is not applicable typically due to the composed structure of the seismic lithosphere zones) should be applied. On the basis of the geological, geophysical, and physical investigations mentioned, the computational stage of construction of the model begins: writing the computational program, its testing, and numerically solving the discretized model, visualization of the numerical results, their analysis, including comparison with the observations and measurements, and recommendations for EQ prognosis.

The authors are not able to perform these reasonable stages (conditions) of the "well known mechanism of the precursory signal generation" for everyone of the precursory phenomena mentioned above. We have chosen (see Chaps. 1, 2, and 3) low-resistivity structures (LRS) as the particularity of a seismic lithosphere zone's structure and EM signals, by restricted consideration of T signals.

1.2 Typicality of Low-Resistivity (0.1–0.5 S/m) Domains for Seismically Active Lithospheric Zones

The typicality of the low-resistivity (0.1–0.5 S/m) domains for seismically active lithospheric zones, along with their characteristic depths and sizes, is the geological basis of our mathematical model of seismo-EM emission, which will be described later. In this section, we refer to geophysical investigations by different authors in different tectonic regions illustrating this typicality, as well as physical and geometric properties of seismic (i.e., seismically active) structures. Numerous magnetotelluric soundings (MTS) and soundings with a magneto-hydrodynamic (MHD) generator have discovered, mostly during the last few decades, considerable low-resistivity (i.e., low-ohmic) domains in seismically active lithospheric zones. The characteristic electrical conductivity of similar domains is of the order of 0.1 S/m (there are cases with 0.5 S/m and up to 1.0 S/m in Tasmania) with typical sizes and a depth of the order of 10 km. To cite the illustrating results of geophysical expeditions in known seismic regions, we begin with the Northwestern Himalaya. Arora and Mahashabde (1987) performed MTS with an array of magnetometers on the profile parallel to the main direction (the size of a structure is maximal in this direction) of this mountain system. The processing of these measurements showed that the domain with electrical conductivity of 0.5 S/m and a width of 45 km, located at depths from 15 km up to the asthenosphere, was in good agreement with the MTS data obtained. The authors noted that (1) their data supported, on the basis of more detailed MTS investigations, the results of Lilley et al. (1981) obtained in Northwestern India in 1979; and (2) the described model of the electro-conductive domain was approximately similar to the model studied by Vozoff (1984). Arora and Mahashabde (1987) supposed that the considered electro-conductive structure extends through the basin of Gang to the foothills of the Himalaya and may represent the zone of stresses created by interaction of the Indian plate with Euro-Asia; these stresses results in cracking in the domain of the Indian shelf and a rising of magma from the asthenosphere.

The next example pertains to the seismic generation zone of North Tien-Shan (including the city of Bishkek), which is known as one of the most active seismic regions of Central Asia. The model of electrical conductivity for this region was constructed by Bragin et al. (1993) on the basis of MTS and deep MTS (DMTS) data and EM sounding with an artificial source. Also, an MHD generator was applied with an impulse system, which was able to create extremely strong current pulses with amplitude values up to 2.5 kA and durations up to 10 s, as well as, for example, a series of stabilized pulses of any desirable duration from 0.1 up to 100 s, with the amplitude value of the current being up to 670 A. According to the authors, the model of electrical conductivity of the earth's crust in the eastern part of the Kyrgyz Tien-Shan includes:

1. An upper local conductive horizon with a mean depth of 10 km, with integral electrical conductivity up to 2700 S and a characteristic capacity of the layer of 5 km
2. An ubiquitous conductive horizon in the lower part of the crust, with the integral electrical conductivity of this horizon increasing from 400 to 2000 S from north to south, i.e., along the line of contact of the southern blocs with the northern ones, with a characteristic thickness of 15 km

Moreover, an especially sharp contrast in the electrical conductivity of the crystalline horizon is observed at the boundary between the main structures. It is necessary to note that a similar intrusive mechanism of formation of the conductive lithospheric structures may be applied for the Northwestern Himalaya. Also, low-resistivity structures have been discovered in the known seismically active region of the San Andreas fault. The data from field investigations in Parkfield, California, were described by Park and Fitterman (1990). According to these data, the authors formulated a model demonstrating that the electrical conductivity of rocks near the San Andreas fault and to the southwest is more than 0.1 S/m up to a depth of 10 km minimum. The authors noted that these low-resistivity structures were created in the course of brecciation (crushing) caused by intensive shifts along the fault. Eberhart-Philips et al. (1990) elaborated the seismic and magneto-telluric models of the epicenter zone of the Loma Prieta earthquake ($M_s = 7.1$, October 17, 1989, Santa Cruz Ridge, California). These models coincide very closely and indicate the existence of a wedge of low-speed and low-ohmic (0.1–0.3 S/m) material near the San Andreas fault. This wedge covers depths up to 8 km. From our viewpoint, the most distinctive large-scale demonstration of correspondence between tectonic processes and structures, on the one hand, and low-resistivity zones, on the other hand, were obtained by investigation of the electrical conductivity of the Australian continent. F. H. Chamalaun and C. E. Barton (1993) used data from a magnetometer array, with a typical step of 275 km, and covered the continent as a whole. They showed that there exists a large U-shaped area of anomalous electrical conductivity, which stretches across the whole country and is localized near the boundaries of the main craton blocs (the term *craton* is used to distinguish the stable portion of the continental crust from regions that are geologically more active and unstable) (Wikipedia Contributors 2018). The electrical conductivity of different parts of the conductive zone described above may be rather high. MTS observations in Eastern Tasmania discovered a body with conductivity of 0.2 S/m at depths from 2 to 5 km (Parkinson et al. 1988). According to these observations, the domain of anomalous conductivity coincides with the Tamar lineament (in Tasmania) and there are geological arguments supporting the supposition about considerable tectonic activity able to cause intensive shifts and cracking of rocks at depths of the order of 2–5 km. Also, F.E.M. Lilley outlined the coincidence of zones of anomalous conductivity and continental seismic zones in Australia (Lilley and Smylie 1968; Lilley 1975). In the case of the continental seismic regions of China, the earthquake epicenters, especially those of the main shocks, were localized above the low-ohmic (from a few up to 20 Ohm/m with a capacity of a few kilometers), low-speed (partially melted rock) layers (Liu Guodong 1991).

The results of MTS, geothermal, and seismological investigations illustrating the connection between high seismicity, electrical conductivity, and heat flow may be found in different geophysical works devoted to the seismic region of Northwestern America; see for example the paper by Gough (1974). A coincidence of zones of high seismicity and ones of high electrical conductivity was discovered by geophysical methods (MTS, MHD). This coincidence is explained, in geophysical works and in the geological theory of ore genesis, by intensive heat transfer and mass transfer along numerous canals of fragmented (because of an active tectonic regime) seismic zones, in particular by intrusion of ultrabasic magmas with Fe, Mn, and other metals in the middle and upper floors of the crust. The extremely nonuniform distribution of the heat flow in the area, in particular an abnormally high heat flow typical for different seismic regions, may be considered a manifestation of these processes of genesis of metals; in other words, the geothermal deposits are located, as a rule, in zones of high seismicity. Confirmation of the above estimations of depths, geometric sizes, and electrical conductivities of seismic structures may be found in Honkura (1974).

We shall not try to include the full data on the petro-physics (Gueguen and Palciauscas 1994) of the low-resistivity rocks of tectonically active lithosphere zones (these data have not been detailed recently) because only the depths, sizes, and conductivities of the structures are used below to model seismo-EM interaction, and these characteristics are known from publications by different authors, mostly from MTS investigations, elastic full-waveform inversion, and other methods of solving inverse problems (Nabighian and Macnae 2005; Raknes and Arnsten 2017). Let us note, nevertheless, that the electrical conductivity of rocks of the order of 0.1 S/m (it is a typical value used in the calculations below) may be provided by the existence of a small percentage of the volume of the low-ohmic ferromagnetic minerals (magnetite, pirotin) or copper–pyrite ores by the condition of galvanic connection of conductive inclusions. Graphitization and mineralized fluids are the usual reasons for similar connections. It will be seen from the further development of the theme that the EM response of rocks on nonstationary deformations (i.e., the main property that is of interest in our seismo-EM problems) increases not only along with an increase in their conductivity but also along with an increase in their magnetic susceptibility. But the electrical conductivity of widespread rocks belongs to the interval of values of a few orders, while their magnetic susceptibility changes by just one order up to 20 in the case of strong magnetics. However, due to the geothermal gradient, at the typical depths of occurrence of large low-ohmic irregularities (a characteristic size is of the order of 10 km at a depth of 20 km or more) of seismic lithosphere zones, the magnetic structures (oriented according to previous magnetization) of the conductors may be regarded as destroyed because the temperature is near the Curie point, whereas the electrical conductivity is preserved. The example considered above and other examples of the distinctive geographical correlation between zones of high seismicity and high electrical conductivity show that it is rational to consider low-ohmic structures of seismic lithosphere zones as the transformers of seismic energy in disturbances of EM fields.

1.3 Observations of EM and Temperature Signals of Seismic Origin

Below, we numerically model ultralow frequency (ULF) EM signals (0.01–10 Hz), which are fixed before the seismic wave arrival (i.e., precursors), and we are going to describe observations of similar signals now. This type of seismo-EM phenomenon is relatively simple from the physical viewpoint: the seismic waves arriving with a definite delay after the ULF signal confirm the hypothesis about the lithospheric origin of the EM signal. That is the reason to choose this type of signal as the first object of the numerical modeling.

To describe some observations of the ULF EM signals prior to a seismic wave, let us cite the following papers (our comments are in brackets).

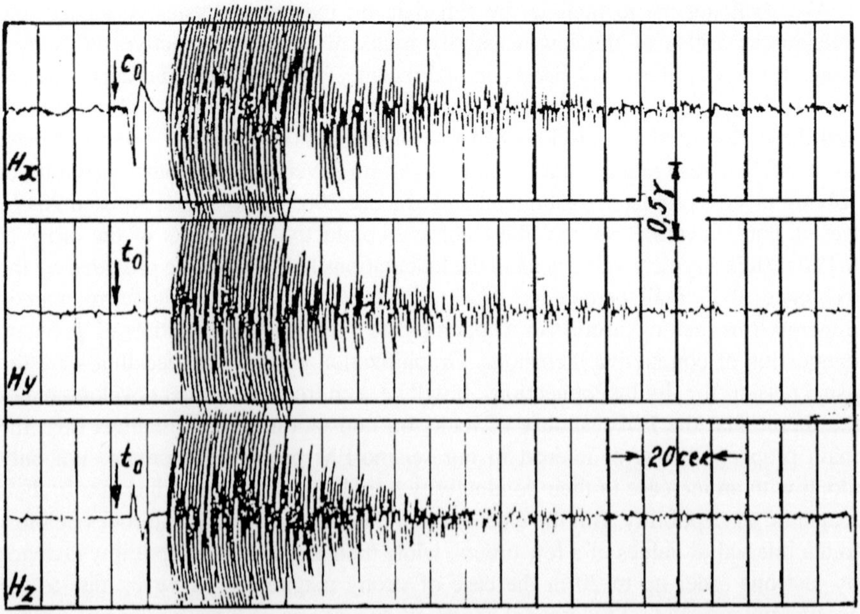

Fig. 1.1 Magnetic observation in Petropavlovsk-Kamchatsky; $M = 6$ (Belov et al. 1974). H_x, H_y denote the north–south (NS) and east–west (OW) horizontal components; H_z denotes the vertical component

Figures 1.1 and 1.2 demonstrate the three-component magnetic records for two earthquakes near Petropavlovsk-Kamchatsky (Belov et al. 1974). The records on December 25, 1972, of an earthquake with magnitude $M = 6$, depth $h = 70$ km, and epicenter distance $R = 70$ km, are shown in Fig. 1.1. Figure 1.2 shows the records on October 11, 1972, with $M = 5$, $h = 80$ km, $R = 50$ km. In both figures, t_0 denotes the moment of the beginning of the earthquake.

Static magnetometers with a photoamplifier were used. The frequency characteristic of the measuring system was flat for the range from 0.1 to 10 Hz. (Let us note that the magnetic induction is denoted in other sections of this work by B). It is seen that the ULF signal was registered before the seismic wave arrival caused the mechanical oscillations of the device as a whole and the geomagnetic field induction in it. According to the recordings, the remote magnetometers in Yuhjno-Sakhalinsk did not register any abnormal atmospheric disturbances on December 25, 1972. Therefore, the registered magnetic effects must have been local and may not have been explained by the influence of atmosphere sources. The authors investigated industrial noise in the region of observations and signals similar to the observed ones in the geophysical research were not discovered. Besides, disturbances on the seismograms, which may be interpreted as the reasons for the shaking of the magnetometers in the moment of registration of the magnetic signal before the seismic wave arrival, were absent. Belov et al. (1974) denote the moment of the beginning of the magnetic signal as t_α and give the following values (see

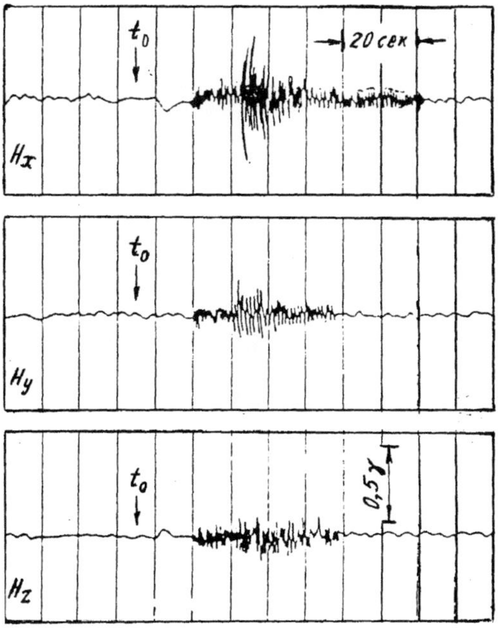

Fig. 1.2 Magnetic observation in Petropavlovsk-Kamchatsky; $M = 5$ (Belov et al. 1974). The notations are the same as those for Fig. 1.1

Figs. 1.1 and 1.2): $t_\alpha - t_0 = 4$ s, $t_p - t_\alpha = 10$ s on December 25; $t_\alpha - t_0 = 5$ s, $t_p - t_\alpha = 9$ s on October 11, where t_p denotes the moment of the seismic wave arrival and t_0 denotes the moment of the beginning of the earthquake. The authors of the paper note that the time intervals $t_\alpha - t_0$ and $t_p - t_\alpha$ are in good agreement for the two considered earthquakes and, probably this is not accidental, because these earthquakes occurred in the same region and were observed at the same point. Let us mention other measurements described by Belov et al. (1974), namely the electro-telluro-gram of the components E_x and E_y of the electrical field recorded at the same time as the seismogram and demonstrating the arrival of the electrical signal 2 s before the seismic wave arrival. Fuginawa et al. (2000) and Takahashi et al. (2000) applied the casing pipe of a deep borehole as a receiving antenna to record the electrical field. A distinctive correlation between abnormal changes of the electrical field and seismic activity was demonstrated.

Gohberg et al. (1991) show a magnetic ULF signal (Fig. 1.3a) prior to the seismic wave (Fig. 1.3b) for an earthquake on September 1, 1988, in Kamchatka; $M = 5.5$, $h = 30$ km, $R = 220$ km. In this case a gradient setup with the base of 1 km was used to reduce the influence of the ionosphere and magnetosphere disturbances. The magnetic field sensors had sensitivity of 0.5×10^{-3} nT, the maximal difference of the phase and frequency characteristics between sensors was no more than 0.1%, and recordings were made in the range from 0.01 Hz up to 4 Hz. It can be seen that the beginning of the oscillation of the magnetic field with the large amplitude coincides with the seismic wave arrival (Fig. 1.3b) recorded by the seismographs at the Kosyrevsk seismic station (in the central domain of Kamchatka) located near the

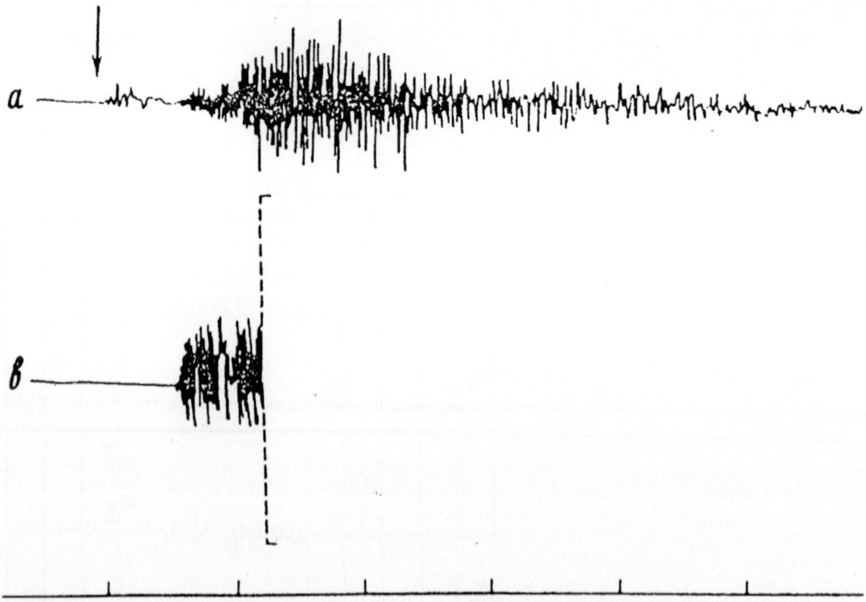

Fig. 1.3 Magnetic (**a**) and seismic (**b**) recordings, Kosyrevsk, Kamchatka; $M = 5.5$, and the interval on the horizontal time scale axis is 1 min (Gohberg et al. 1991)

place of observation of the magnetic signals. The magnetic signal (Fig. 1.3a) prior to the seismic wave arrival has a spectrum without high-frequency components, contrary to the initial interval of the seismic recordings in Fig. 1.3b. The observable period of the zug of the unidentified magnetic oscillations is from 2 to 6 s, whereas the amplitude of the signal in the difference canal is from 2 to 10 nT with a duration $\tau = 30$ s. Records from the standard seismograph (Kosyrevsk) do not demonstrate any seismic disturbance at the moment of the magnetic ULF signals [similarly to the case described by Belov et al. (1974)]. Besides, the general background of EM disturbances during the period of the observations (September 1988) was quiet. Gohberg et al. (1991) suppose that the EM signal (Fig. 1.3a) was caused by the earthquake. As the reasons for this conclusion they note that (a) the beginning of this signal practically coincides with the moment t_0 of the beginning of the earthquake (the delay of the signal in regard to the earthquake beginning is seen, nevertheless); and (b) high-frequency components are absent in the signal under consideration (possibly because of attenuation in conductive rocks).

Dea et al. (1994) describe a "large zug" of EM impulses 15 s (approximately) before the shock of the Big Bear earthquake on June 28, 1992 ($M = 6$, Southern California, $R = 150$ km approximately; EM signals in the range of 0.1–20 Hz were recorded). "One of us (J. Dea) was at the observation station in order to register any disturbances caused by the Landers earthquake [which] occurred 3 h before the Big Bear earthquake.... The automatic registration with averaging and recordings for every 15 min occurred to be insufficient to register this group of waves." The group of waves was not registered but the example of a signal was demonstrated with a similar form (this signal was distinctive enough in the general background); the number of oscillations in this signal far exceeded the number of oscillations in the signals recorded by the authors, as mentioned above.

As an addition to the above citations of the authors observing an EM signal before the seismic wave arrival, it ought to be noted that in one of the cases described by Belov et al. (1974), the time delay was $t_\alpha - t_0 = 5$ s (see above). This delay may have been about 10 s in the case of Dea et al. (1994) because $t_\alpha - t_0$ depends on h, R, the depth of conductive structures, and the trace of the signal in the air; these conditions are different in the cases of Dea et al. (R = 150 km) and Belov et al. (R = 50 km).

Therefore, the time of propagation of the seismic wave in the case described by Dea et al. (1994) may have been of the order of 25 s and its velocity may have been of the order of 6 km/s; this is a typical value. The numerical illustration of a delay in an EM signal in regard to the moment of the earthquake beginning will be considered below.

So, according to the cited observations we shall consider, as a numerical example, the deformation caused by a seismic wave for the case of a model lithosphere domain with low-resistivity inclusions.

Observations of the EM field along with the temperature (T) field in seismically active regions are of importance because they are convincing evidence of the prognostic value of changes in these fields. Some illustrations of the EM field were given above. Let us look at observations of the T field caused by seismic events.

At the start, it is necessary to note that the T changes are not connected with the dynamics of the ionosphere–magnetosphere plasma, contrary to the EM field. Therefore, geothermal investigations—say, with quartz thermo-resistors in boreholes—may be used to support the supposition of the lithospheric origin of the measured ULF EM signals accompanied by the T disturbances. On the other hand, absence of T disturbances may be considered an argument for discrimination of the EM field changes as ones of nonlithospheric origin when prognostic problems are concerned. We believe it is reasonable to suppose that the surface T oscillations with their diurnal and seasonal variations are spectrally different from the T changes caused by lithosphere dynamics. These differences may be specific for the given seismic region and must be registered repeatedly for prognostic applications. Moreover, these differences may change after seismic events and must be checked.

According to Mei Shirong (1992), a digital precision quartz thermometer with temperature resolution of 0.0001 °C was applied in China in 1983 to register the geothermal precursors of earthquakes. The geothermic stations in different regions of China accumulated hundreds of thousands of hours of geothermic measurements during the years 1987–1989. The time intervals of geothermal precursors may be registered only in the neighborhood of the epicenter of a future earthquake and the radius of this neighborhood is proportional to the magnitude of the expected earthquake. To monitor earthquakes with M = 5–6, the stations must be located at a distance of 70–100 km from each other, whereas for the case of M = 7 this distance may be 100–200 km. The variations of the heat flow and groundwater levels before and during an earthquake were measured in seismic regions of

Caucasus (http://www.gbv.de/dms/clausthal/E_BOOKS/2011/2011EB1129.pdf 05.01.2019) and Central Asia (Kissin et al. 1996). The typical temperature disturbances at distances up to 100 km from the epicenter of a future earthquake during the time interval from a few days up to a few months before the earthquake were 0.01–1.0 °C. A review of observations of changes of temperature in connection with seismic events on the basis of world data may be found in Zubkov (1992). According to this author, the area of propagation of thermal precursors is not more than the area of the deformation precursors and these precursors arise almost at the same time as hydro-geodynamic ones, i.e., after 0.9 of the time of the preparation of an earthquake. These precursors may be connected with changing conditions of heat and mass transfer because of arising of cracks. Seismically originated ionospheric variations were described by Sarkar et al. (2007). Moreover, groundwater levels and temperature are observed usually in connection with the earthquake preparation processes (Orihara et al. 2014).

We believe that the cited works are sufficient for a preliminary perspective on earthquake precursors and to get input data for their numerical modeling.

1.4 Remarks About Inverse Problems

The main goal of the geophysical part of this book is quantitative investigation of the physical processes in zones of preparation of earthquakes, i.e., interaction of geophysical fields of different natures in these zones. The geological and geophysical properties of these zones determine the development of the geophysical field interaction processes. The problem of estimation of the physical and geometric parameters of these dipped media and their changes, by the accordance between the measured data and the calculated medium's model (taking account of restrictions of the prospecting signal generation and the measuring devices available) is of the most theoretical and practical importance in physics, biophysics, geology, geophysics, etc.

This class of physical problems is known as inverse problems (to determine the characteristics of a medium by knowing about the physical fields in it), in contrast to direct problems (to determine the physical fields in a medium by knowing about its characteristics).

In a typical situation, a researcher sends one or a few investigating (prospecting) signals of different physical natures in the medium under investigation and records the medium's reaction, i.e., the fields generated in it. Usually a restricted set of the points and/or moments of the time where and when the measuring devices may be applied is available for observation. These measured data occur to be known in a restricted number of points—spatial and/or temporal ones—with a considerable error of 10–20% according to the field conditions of the measurements.

Let us suppose that some model of the medium under investigation is determined and it is necessary to establish whether this model is admissible. Typically, there exist some restrictions in the physical and geometric parameters of the medium that are known before the physical investigation (e.g., from the physics of the rocks, i.e., petrophysics) and an admissible model must obey these preliminary restrictions. Surely, an admissible model must obey not only the preliminary restrictions but the results of the measurements as well: its reaction to the investigating signals must be similar to the measured reaction of the real investigated medium, i.e., the difference between the model and the observed reactions must be no more than the error of the field measurements. So, the model of the medium is considered to be admissible if the preliminary and measured restrictions are obeyed by this model.

We must note that quite often the goal of the geological interpretation of the measured data is considered as follows: to find out the model of the medium (let us denote this model as the model*) for which the reaction to the investigating (i.e., prospecting) signal is the nearest one to the measured reaction, as compared with reactions of other admissible models. But it is not sufficient to find the model* because other admissible models~ of the medium may exist that are more perspective from the point of view of productivity then model* and the solution about drilling (let us suppose that oil, gas, or industrial waters are meant) is connected closely with models~. Thus, the problem of massiveness of the set of admissible models arises.

Let us suppose the set of admissible models is not empty (it is a separate problem of compatibility of the preliminary information about the medium and measurements). Then the question arises about determining something like the massiveness (*cardinality* is a mathematical term from the theory of sets; it is not the most proper one here) of the set of admissible models—say, its diameter in the sense of the metrics in the space of the models, or Kolmogorov's diameter, etc. The complex of measurements and interpreting calculations is supposed to be effective if the set of admissible models is decreasing up to the needed degree of accuracy in the course of increasing the number of the measurements. There are several difficult problems of the theory of inverse problems which were not listen there, e.g., is it possible to miss an interesting (from the point of view of productivity) variant of an admissible model structure by regularization of the inverse problem under consideration?

The more precise formulation of the inverse problems in terms of the functional spaces, the diameters of their subsets, the theory of optimization, including the optimal control and the mathematical programming for the nonlinear inverse problems with a finite number of variables, the functionals of quality of interpretation, and its stability in regard to the restricted accuracy of the measurements, are omitted here.

Inverse problems are discussed in various works (Tarantola 1987; Korovkin et al. 2007; Kapio and Somersalo 2010; Special Section on Inverse Problems in Geosciences 2013; Knopov and Kasitskaya 2002; Kryzhanovsky et al. 2017; Menendez et al. 2000; Menke 2018; Raknes and Arnsten 2017; Sochelnikov 2006; Strakhov 2008) and in academic journals such as *Inverse Problems in Science and Engineering*, *Journal of Inverse and Ill-Posed Problems*, and *Inverse Problems and Imaging*. Computed tomography (CT), with its ability to get images of the sequential cross-sections of a three-dimensional (3D) body, using an external source of investigating rays and processing with integral equations of type I, Radon transform etc., is a specific kind of inverse problem applied in brain, cardiac, and industrial research (Kak and Slaney 1987).

So, we shall not consider the inverse problems theory in detail in a book about the interacting dynamical geophysical fields of different natures, precursors to land and marine earthquakes, tsunamis, and the impacts on the brain and heart, though the field interaction depends on the characteristics of the medium where the fields are localized, and these characteristics may be estimated in the frames of inverse problems, computational experiments, and mathematical theory.

References

Arora, B. R., & Mahashabde, M. V. (1987). A transverse conductive structure in the Northwest Himalaya. *Physics of the Earth and Planetary Interiors, 45*, 119–127.

Belov, S. V., Migunov, N. I., & Sobolev, G. A. (1974). Magnetic effects of strong earthquakes on Kamchatka. *Magnetism and Aeronomy, 14*, 380–382. (In Russian: Белов С.В., Мигунов Н.И., Соболев Г.А. Магнитные эффекты сильных землетрясений на Камчатке. Магнетизм и аэрономия. 1974, 14, с. 380–382).

Bragin, V. D., Volykhin, A. M., Trapesnikov, Y. A., Koshkin, N. A., & Batalev, V. Y. (1993). Depth structure of seismicity-dangerous zones of the Tien-Shan based on electromagnetic sounding data. *Journal of Earthquake Prediction Research, 2*, 329–338.

Chamalaun, F. H., & Barton, C. E. (1993). The large scale electrical conductivity structure of Australia. *Journal of Geomagnetism and Geoelectricity, 45*, 1209–1212.

Dea, J. Y., Hansen, P. M., & Boerner, W. M. (1994). Direct observation of real time generation of electromagnetic signals and a generalized model for earthquake precursor emission. In M. Hayakawa & Y. Fujinawa (Eds.), *Electromagnetic phenomena related to earthquake prediction* (pp. 439–449). Tokyo: Terra Scientific (Terrapub).

Eberhart-Phillips, D., Labson, V. F., Stanley, W. D., Michael, A. J., & Rodriguez, B. D. (1990). Preliminary velocity and resistivity models of the Loma Prieta earthquake region. *Geophysical Research Letters, 17*, 1235–1238.

Fuginawa, Y., Takahashi, K., Matsumoto, T., Iitaka, H., Yamane, S., Nakayama, T., et al. (2000). Electromagnetic field anomaly associated with the 1998 seismic swarms in central Japan. *Physics and Chemistry of the Earth, Part A, 25*, 247–253.

Gohberg, M. B., et al. (1991). Electromagnetic field of an earthquake's focus. *Proceedings of RAS, 308*(1), 62–65. (In Russian: Гохберг М.Б., Крылов С.М., Левшенко В.Т. Электромагнитное поле очага землетрясений. Доклады АН СССР. 1991, 308, № 1, с. 62–65).

Gough, D. I. (1974). Electrical conductivity under western North America in relation to heat flow, seismology and structure. *Journal of Geomagnetism and Geoelectricity, 26*, 105–123.

Gueguen, Y., & Palciauscas, V. (1994). *Introduction to physics of rocks*. Princeton: Princeton University Press.

Guodong, L. (1991). *MT studies on the conductivity of the crust and upper mantle in China*. Beijing: Annual Review in Geophysics Seismological Press.

Honkura, Y. (1974). Electrical conductivity anomalies beneath the Japan arc. *Journal of Geomagnetism and Geoelectricity, 26*, 147–171.

Kak, A. G., & Slaney, M. (1987). *Principles of computerized tomographic imaging*. New Jersey: IEEE.

Kapio, J., & Somersalo, E. (2010). *Statistical and computational inverse problems*. New York: Springer.

Kissin, I. G., Belikov, V. M., & Ishankuliev, G. A. (1996). Short-term groundwater level variations in a seismic region as an indicator of the geodynamic regime. *Tectonophysics, 265*(3–4), 313–326.

Knopov, P., & Kasitskaya, E. (2002). *Empirical estimates in stochastic optimization and identification*. New York: Springer.
Korovkin, N. V., Chechurin, V. L., & Hayakawa, M. (2007). *Inverse problems in electric circuits and electromagnetics*. New York: Springer.
Kryzhanovsky, B., et al. (Eds.). (2017). *Advances in neural computations, machine learning, and cognitive research* (p. 199). New York: Springer.
Lilley, F. E. M. (1975). Electrical conductivity anomalies and continental seismicity in Australia. *Nature, 257*, 381–382.
Lilley, F. E. M., & Smylie, D. E. (1968). Elastic wave motion and a nonuniform magnetic field in electrical conductors. *Journal of Geophysical Research., 73*(20), 6527–6533.
Lilley, F. E. M., Singh, B. P., Arora, B. R., Srivastava, B. I., Plasad, S. N., & Sloan, M. N. (1981). A magnetometer array study in northwest India. *Physics of the Earth and Planetary Interiors, 25*, 232–240.
Menendez, R. G. P., Andino, S. L. G., Morand, S., Michel, C. M., & Landis, T. (2000). Imaging the electrical activity of the brain: ELECTRA. *Human Brain Mapping, 9*, 1–12.
Menke, W. (2018). *Geophysical data analysis (Discrete inverse theory)* (4th ed.). Amsterdam: Elsevier.
Nabighian, M. N., & Macnae, J. C. (2005). Electric and EM methods, 1980–2005. *The Leading Edge, 24*, 542–545.
Orihara, Y., Kamigawa, M., & Nagao, T. (2014). Preseismic changes of the level and temperature of confined groundwater related to the 2011 Tohoku earthquake. *Scientific Reports, 4*, 6907.
Park, S. K., & Fitterman, D. V. (1990). Sensitivity of the telluric monitoring array in Parkfield, California, to changes of resistivity. *Journal of Geophysical Research, 95*(15), 557–571.
Parkinson, W. D., Hermanlto, R., Sayers, J., Bindoff, N. L., Dosso, H. V., & Nienaber, W. (1988). The Tamar conductivity anomaly. *Physics of the Earth Planetary Interiors, 52*, 8–22.
Raknes, E. B., & Arnsten, B. (2017). Challenges and solutions for performing 3D time-domain elastic full-waveform inversion. *The Leading Edge, 36*(1), 88–93. https://doi.org/10.1190/tle36010088.1. Accessed 10 January 2019.
Sarkar, S., Gwal, A. K., & Parrot, M. (2007). Ionospheric variations observed by the Demeter satellite in the mid-latitude region during strong earthquakes. *Journal of Atmospheric and Solar-Terrestrial Physics, 69*, 1524–1540.
Shirong., M. (1992). Progress in earthquake prediction in China during 80's. *Journal of Earthquake Prediction Research, 1*, 43–57.
Sochelnikov, V. V. (2006). *The high-resolution electro-prospecting*. (In Russian: Высокоразрешающая электроразведка. Морская Государственная Академия имени адмирала Ф.Ф.Ушакова).
Special Section on Inverse Problems in Geosciences. (2013). *Inverse Problems in Science and Engineering 21*(3), pp. 355–561.
Strakhov, V. N. (2008). The list of publications. In: *My life in science* (Vol. 1, pp. 19–67). Moscow: Schmidt Institute of Physics of the Earth, RAS. (In Russian: Список публикаций. Моя жизнь в науке. Т. 1. ИФЗ РАН. С. 19–67).
Takahashi K., Fujinawa Y., Matsumoto T., Nakayama T., Sawada T., Sakai H., et al. (2000). *An anomalous electric field variation associated with the seismic swarm (1)-underground electric observation at Hodaka station*. Technical note of the National Research Institute for Earth Science and Disaster Prevention, No. 204, pp. 1–204.
Tarantola, A. (1987). *Inverse problem theory: Methods for data fitting and model parameter estimation* (p. 630). Amsterdam: Elsevier.
Vozoff, K. (1984). Model study for the proposed magnetotelluric (MT) traverse in North India. *Tectonophysics, 105*, 399–411.
Wikipedia Contributors. (2018). *Craton. Wikipedia, the free Encyclopedia*. Accessed October 29, 2018, from https://en.wikipedia.org/w/index.php?title=Craton&oldid=863967996
Zubkov, S. I. (1992). Thermal precursors of earthquakes. *Physics of the Earth, 8*, 72–82. (In Russian: Зубков С.И. Термические предвестники землетрясений. Физика Земли. 1992, № 8, с. 72–82).

Chapter 2
Theory of Magnetic Location of a Possible Earthquake Epicenter Area

Abstract In this chapter, on the basis of the three-dimensional (3D) dynamic theory of magneto-thermo-elasticity and geological and geophysical input data, the theory of magnetic location of a possible earthquake epicenter area (MLPEEA) has been formulated for the first time and compared (satisfactorily) with the results of electromagnetic (EM) measurements in seismically active regions, published by other authors.

MLPEEA is understood here as an area with measurable repeated emissions of a hertz-range EM field caused by relatively weak (maybe regarded as precursory) seismic excitations.

Because of the different (mechanical and EM) natures of the physical fields involved, the mathematical types of the differential operators included in the seismo-EM interaction model are different as well (hyperbolic and parabolic), and the resulting model is not of a well-investigated type. In our previous works, we mathematically investigated the correctness of the model for the continuous and discretized variables, and we have applied these results here for construction of the model of the magnetic location mentioned.

Practical recommendations for earthquake monitoring are given with application of the developed theory. To demonstrate the stability of the effect of the location in regard to the inevitable uncertainty of the geological input data, we carried out simulations of the magnetic location with different reasonable (from the point of view of geophysics and petrophysics) values of the input parameters. In the course of these computations we observed imprinting of the seismic wave on the EM wave, and we describe this phenomenon in the last part of the chapter.

Keywords Seismo-EM interaction · Epicenter area · Magnetic location · Mathematical model · String–diffusion

O. Novik et al., *Electromagnetic Geophysical Fields*,
https://doi.org/10.1007/978-3-319-98461-2_2

2.1 Geological and Physical Basis of the Theory

According to the results of numerous geophysical field investigations (mostly magneto-telluric soundings) and the geological theory of ore genesis (e.g., Robb 2005), high-seismicity lithosphere zones typically contain high-electrical-conductivity (of the order of 0.1 S/m) structures. Taking account of this particularity of seismically active lithosphere zones, we apply the physical theory of magneto-thermo-elasticity to formulate a mathematical model of the seismo-electromagnetic (seismo-EM) phenomenon of magnetic location of an area emitting a seismically generated EM field, repeated emissions being an important sign of the possibility of an earthquake; namely, by one of the reasonable prognostic interpretations, a similar area is expected to be the area of the epicenter of a future earthquake, and its coordinates and azimuths, from different observation points, may be ones of a forthcoming epicenter area. Practical recommendations for earthquake monitoring are given with application of the developed theory of magnetic location of the epicenter area.

By magnetic location of a possible earthquake epicenter area (MLPEEA), a geometric image of the measured magnetic field is constructed (e.g., a set of some arrows or perpendiculars to some surfaces, etc., connected with the field; see below) to estimate the azimuths and/or horizontal coordinates of the seismically activated dipped source of the emitted field. The projection of the source of the emission on the earth's surface may pertain to the epicenter area of a future earthquake.

MLPEEA was projected and accomplished for the first time by Kopytenko and co-authors (Kopytenko et al. 1999, 2000; Ismaguilov et al. 2001, 2003). Our theoretical results correspond to these observations, i.e., they demonstrate the picture of MLPEEA.

The physical nature of MLPEEA is not known well. We suggest a mathematical model of MLPEEA on the basis of the geological theory of ore genesis and the physical theory of magneto-thermo-elasticity.

The geological component of our model proceeds from the known geographic correlation between lithosphere zones with high seismicity and ones with low-resistivity structures. The conductivity of these structures is of the order of 0.1 S/m with characteristic size and depth of the order of 10 km. In the geological theory of ore genesis, this correlation is explained by the existence of plenty of channels for upward motion of fluids with melted metals from the asthenosphere to the crust, and these channels are considered to be a result of fragmentation of a geological medium due to an intensive tectonic regime (Arora and Mahashabde 1987).

The magneto-thermo-elasticity theory (Maugin 1988) is used to describe evolution of seismic disturbances of the geophysical fields mentioned. We suppose that relatively weak (i.e., elastic) seismic waves arise during an earthquake preparation process as local decreases of stress, because monotonic growth of deformations in all parts of a seismic preparation area, with a typical size of the order of 1000 km, is hardly possible. As a result, an EM field is being generated by a similar precursory seismic wave deforming a low-resistivity structure (the latter may be considered a seismo-EM transformer) in the presence of the geomagnetic field. Besides, the rheological stress–strain relation of a rock may include EM and thermal terms (dissipation).

So, we believe that precursors to earthquakes are being formed by a seismo-EM-thermal interaction (SEMTI) process in the lithosphere zones with the low-resistivity seismo-EM transformers typical of similar zones (Novik 1969, 1994, 1995; Novik and Ershov 2001). In these publications a mathematical theory of the nonclassical string–diffusion systems of partial differential equations (PDEs), describing, in particular, the magneto-thermo-elasticity interaction, was developed and seismo-EM transformation was described quantitatively, i.e., interaction of elastic deformations (string) and EM diffusive processes.

The EM processes are considered by diffusion approximation, because the displacement currents are negligible for the low-frequency (0.01 ÷ 100 Hz) processes in a conductive medium (0.001 ÷ 0.1 S/m). So, we will formulate and investigate the MLPEEA mathematical model, including the comparison with observations, in this framework.

As far as we know, a mathematical model of the seismo-EM phenomenon of MLPEEA on the basis of the geological theory of ore genesis, the physical theory of magneto-thermo-elasticity, and mathematical investigation of the nonclassical string–diffusion systems of PDEs, is new.

2.2 String–Diffusion Mathematical Model of the SEMTI Process

2.2.1 The Model Medium

The SEMTI process will be considered in the model lithosphere medium (domain) R_l:

$$R_l = \{(x, y, z); 0 < x < L_x, 0 < y < L_y, 0 < z < L_z\},$$
$$L_x = L_y = 160, L_z = 40 \text{ (the upper mantle), km.}$$

The lithosphere domain R_l includes (see Fig. 2.1):

- The low-resistivity block E (seismo-EM transformer) in the form of the ellipsoid,

$$E = \left\{(x, y, z) : \left(\frac{x - x_0}{l_x}\right)^2 + \left(\frac{y - y_0}{l_y}\right)^2 + \left(\frac{z - z_0}{l_z}\right)^2 = 1\right\},$$

$l_x = 12.8$, $l_y = 10.4$, $l_z = 5.2$, km, $(x_o, y_o, z_o) = (80, 80, 20)$, km, is in the center of R_l;

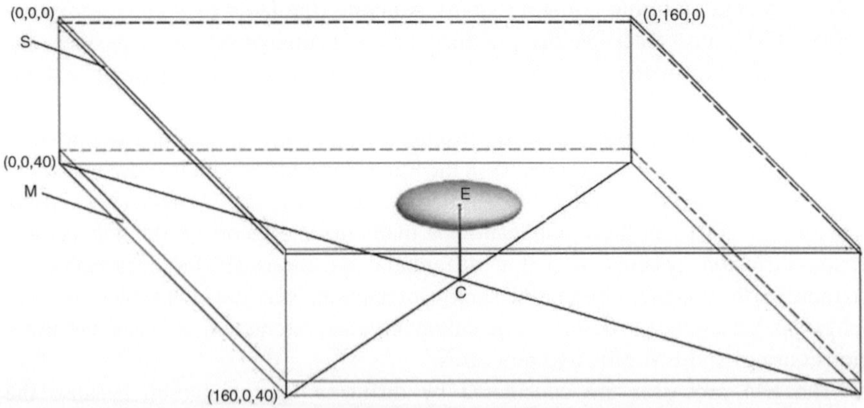

Fig. 2.1 Structure of the model lithosphere domain R_l with the low-resistivity block E, sedimentary layer S, and upper mantle layer M. The drawing to scale is approximate; the geometric characteristics of R_l, E, S, and M are written above. The coordinates and distance between points E and C (20) are in kilometers

Table 2.1 Physical properties of the model medium with the low-resistivity block

	ρ (kg/m^3)	λ (GPa)	μ (GPa)	σ_e (S/m)	k (W/m K)
E	4800	82.8	25.0	0.1	5.0
S	2700	27.5	20.0	0.001	1.3
M	4800	40.0	25.0	0.005	5.0
R_{lc}	2800	51.1	30.0	0.001	2.5

E low-resistivity block, k heat conductivity, λ Lame elastic parameter, M upper mantle layer, μ Lame elastic parameter, ρ density, σ_e electrical conductivity, R_{lc} lithosphere domain (R_l) without E, S, and M, S sedimentary layer

- The sedimentary layer S:

$$S = \{(x, y, z); 0 < x < L_x, 0 < y < L_y, 0 < z < L_S\}, L_S = 1 \text{ km};$$

- The upper mantle layer M:

$$M = \{(x, y, z); 0 < x < L_x, 0 < y < L_y, L_M < z < L_z\}, L_M = 36 \text{ km}.$$

The lithosphere domain R_l without E, S, and M is denoted by $R_{lc} = R_l/(E + S + M)$.

So, the geometric characteristics of the model medium have been defined; let us describe the physical ones.

In the lithosphere domain R_l, the values of the density ρ, Lame elastic parameters λ and μ, electrical conductivity σ_e, and heat conductivity k are determined as shown in Table 2.1.

The values of the physical characteristics of the model medium are usual from the viewpoint of the physics of the rocks (Schön 2011), save for the relatively high electrical conductivity of block E, which is the typical particularity of seismically active lithosphere domains according to field expeditions and the geological theory of ore genesis (see Sect. 2.1).

2.2.2 Field Equations

Let us formulate—with the proper notations for our case—field equations of the magneto-thermo-elasticity (MTE) theory for the SEMTI process in R_1.

Weak seismic excitations (i.e., ones with elastic deformations of rocks) are supposed to supply energy for a SEMTI process of generation of precursory EM signals. Elastic deformations may be described by operator $\mathcal{H}$ of the linear dynamical theory of elasticity:

$$\mathcal{H}\mathbf{u} = \rho\mathbf{g}, \tag{2.1}$$

where:

$$\mathbf{u} = (u_1(t,\mathbf{x}), u_2(t,\mathbf{x}), u_3(t,\mathbf{x}))^{\mathrm{tr}}, \mathbf{x} = (x_1, x_2, x_3) = (x, y, z), \mathbf{g} = (g_1, 0, 0)^{\mathrm{tr}}$$

are the vectors of the elastic displacements, coordinates, and acceleration of gravity, respectively, the indices 1 and 2 are used for the horizontal directions, 3 is for the vertical one, and $\rho = \rho(\mathbf{x})$ denotes the density;

$$\begin{gathered}\mathcal{H}\mathbf{u} \equiv \rho\mathbf{u}_{,tt} - \mathrm{Div}\widehat{\sigma}(\mathbf{u}),\\ \widehat{\sigma} = (\sigma_{ij})_{i,j=1}^{i,j=3},\ \sigma_{ij}(\mathbf{u}) = 2\mu(\mathbf{x})\varepsilon_{ij} + \lambda(\mathbf{x})\delta_{ij}div\mathbf{u},\\ \varepsilon_{ij} = \frac{1}{2}(u_{i,j} + u_{j,i}),\ \ i,j = 1,2,3\end{gathered}$$

We apply the short notation for derivatives:

$$\partial\mathbf{W}/\partial x_i \equiv \mathbf{W}_{,i}, \partial\mathbf{W}/\partial t \equiv \mathbf{W}_{,t},\ \ i = 1,2,3$$

So, the stress tensor σ_{ij} is supposed to obey the Hook strain–stress relation with the Cauchy strain tensor ε_{ij} and Lame elastic parameters μ and λ, and $\mathrm{Div}\widehat{\sigma}$ denotes the divergence of the stress tensor,

$$\mathrm{Div}\widehat{\sigma}(\mathbf{u}) \equiv \mathrm{Div}\widehat{\sigma} \equiv \left(\left(\mathrm{Div}\widehat{\sigma}\right)_1, \left(\mathrm{Div}\widehat{\sigma}\right)_2, \left(\mathrm{Div}\widehat{\sigma}\right)_3\right)^{tr},$$

$$\left(\mathrm{Div}\widehat{\sigma}\right)_i = \sum_{j=1}^{3} \sigma_{ij,j}, \quad i = 1,2,3$$

Operator $\mathcal{H}$ belongs to the hyperbolic type (the operator of a string is the simplest example): the mechanical energy of deformation processes described by similar operators does not disappear. But it is not true for rocks, because temperature and EM fields are being generated by deformations.

In the case of seismic generation of field disturbances in conductive rocks with a minimal electrical conductivity > 0.0001 S/m, these processes include propagation of a low-frequency (from 0.01 up to 100 Hz) diffusive field of magnetic induction **B** $(t, \mathbf{x})$,

$\mathbf{B}(t,\mathbf{x}) = (B_1(t,\mathbf{x}), B_2(t,\mathbf{x}), B_3(t,\mathbf{x}))^{\mathrm{tr}}, (\mu_0)^{-1}\mathbf{B}(t,\mathbf{x}) = \mathbf{H}(t,\mathbf{x}) = (H_1(t,\mathbf{x}), H_2(t,\mathbf{x}), H_3(t,\mathbf{x}))^{\mathrm{tr}}$ is the magnetic field intensity, and the thermal field with deviation $\theta(t, \mathbf{x})$ of the temperature $\Theta(t, \mathrm{x})$ from the initial temperature θ_0 existing at $t = 0$, i.e., before the seismic excitation,

$\mathbf{w}(\mathrm{t}, \mathbf{x}) \equiv (\mathbf{H}(t, \mathbf{x}), \theta(t, \mathbf{x}))^{\mathrm{tr}}$ is the notation used to combine these dissipative processes.

The medium is supposed to be uniform in regard to magnetic permeability, and its value is $\mu_0 = 4\pi \cdot 10^{-7}$ (in SI units).

Besides, let us suppose that the polarization (magnetic and electrical) and thermoelectric processes are negligible, as well as the change in the physical characteristics of rocks under the influence of a weak precursory excitation.

So, diffusion—magnetic and thermal diffusion in the medium under consideration—is described by a system of PDEs of the parabolic type:

$$\begin{aligned} \mathbf{B}_{,t} + \mathrm{rot}\left(\sigma_e^{-1}(\mathbf{x})\mathrm{rot}\mathbf{H}\right) &= 0 \\ \rho(\mathbf{x})c_V\theta_{,t} - \mathrm{div}(\kappa(\mathbf{x})\mathrm{grad}\vartheta) &= q \end{aligned} \tag{2.2}$$

where:

$\mathbf{q} = (0, 0, q(\mathbf{x}))^{tr}$ is the density of power of heat sources.

Here the first line, which is the vector with three components, determines the operator of magnetic diffusion (the low-frequency displacement currents are negligible as compared with ones of conductivity in R_1) by electrical conductivity σ_e and the second line determines the scalar operator of heat transfer (diffusion) by heat conductivity κ; heat capacity is denoted by c_V.

Contrary to observations in seismically active regions, the classical hyperbolic and parabolic vector equations (Eqs. 2.1 and 2.2) are not able to describe generation of magnetic (**H**) and temperature (θ) field disturbances by seismic excitation of a medium, i.e., magnetic and temperature precursors to earthquakes.

By the suppositions on the medium and processes mentioned, we apply the magneto-thermo-elasticity theory (Maugin 1988) to describe seismo-EM-temperature interaction in the lithosphere. Then Eqs. (2.1) and (2.2) are replaced by the coupled string–diffusion system of Eqs. (2.3) and (2.4):

$$\mathcal{H}\mathbf{u} + \mathcal{A}(\mathbf{w}) = \rho\mathbf{g} \tag{2.3}$$

$$\begin{aligned} \mathbf{B}_{,t} + \mathrm{rot}\left(\sigma_{\mathrm{e}}^{-1}(\mathbf{x})\mathrm{rot}\mathbf{H}\right) &= \mathrm{rot}[\mathbf{v} \times \mathbf{B}] \\ \rho(\mathbf{x})c_V\theta_{,t} - \mathrm{div}(\kappa(\mathbf{x})\mathrm{grad}\vartheta) &= q + \mathcal{B}(\mathbf{u}, \mathbf{w}) \end{aligned} \tag{2.4}$$

with the string–diffusion interaction operators:

$$\begin{gathered} \mathcal{A}(\mathbf{w}) \equiv \mathrm{grad}(\beta(\mathbf{x})\vartheta) + \mathbf{B} \times \mathrm{rot}\mathbf{H} \\ \mathrm{rot}[\mathbf{v} \times \mathbf{B}] \\ \mathcal{B}(\mathbf{u}, \mathbf{w}) \equiv -\beta(\mathbf{x})\vartheta_0 \mathrm{div}\mathbf{v} + \sigma_{\mathrm{e}}^{-1}(\mathbf{x})(\mathrm{rot}\mathbf{H})^2 \end{gathered}$$

$\mathbf{v} = \partial\mathbf{u}/\partial t \equiv \mathbf{u}_{,t}$, $\mathbf{v}^2 = v_1^{\,2} + v_2^{\,2} + v_3^{\,2} \ll c^2$, c is the light velocity in a vacuum (this is the usual assumption of "slow motion"); $\beta = (2\,\mu + 3\lambda)\alpha_\theta$, α_θ is the coefficient of thermal expansion, $\alpha_\theta = 5 \times 10^{-6}$ K^{-1}. The first term in the expression for interaction operator $\mathcal{A}(\mathbf{w})$ describes the thermal stress and the second one originates from the Lorentz force. The interaction operator rot[$\mathbf{v} \times \mathbf{B}$] arises by applying *rot* to the Maxwell equation (without displacement currents; see above): rot $\mathbf{H} = \mathbf{j}$, $\mathbf{j} = \sigma_{\mathrm{e}}(\mathbf{E} + [\mathbf{v} \times \mathbf{B}])$, $\mathbf{j}$ is the density of the conductivity current in a "slowly moving" medium; $\mathbf{E}$ is the electrical intensity. The first and the second terms of the operator $\mathcal{B}(\mathbf{u}, \mathbf{w})$ describe heat production because of motion and electrical currents, respectively.

System (2.3), (2.4) of the PDEs of the SEMTI is not of a classical type because of the different, namely string and diffusion, physical natures of processes **u**, **w** and, correspondingly, the different mathematical types of the operators involved (see above).

Let us note that the string–diffusion operators are vanishing by $\mathbf{w} \to 0$, $\mathbf{u} \to 0$. In other words, if the physical fields under consideration are weak enough then the string–diffusion (S.-D.) system of PDEs takes the form of Eqs. (2.1), (2.2) of the classical types.

On the basis of our preliminary calculations (Novik and Ershov 2001), we have concluded that (a) the temperature variations, being informative as the earthquake precursors, do not considerably influence the seismo-EM signals, i.e., our main subject, and we shall calculate seismo-EM interaction without the heat transfer processes; and (b) the terms of the operator A in the equation of motion (2.3) do not influence the motion and, therefore, the other processes.

So, we shall consider the three-dimensional (3D) system of equations of the dynamical theory of elasticity and magnetic diffusion:

$$\begin{gathered} \mathcal{H}\mathbf{u} = \rho\mathbf{g} \\ \mathbf{B}_{,t} + \mathrm{rot}\left(\sigma_e^{-1}(\mathbf{x})\mathrm{rot}\mathbf{H}\right) = \mathrm{rot}[\mathbf{v} \times \mathbf{B}] \end{gathered}$$

Let us separate the initial (stationary)—i.e., before the seismic activation—configurations of the fields and their seismic disturbances: $\mathbf{u} = \mathbf{u}_0 + \mathbf{u}_S$, $\mathbf{H} = \mathbf{H}_0 + \mathbf{H}_S$, $\mathbf{B} = \mathbf{B}_0 + \mathbf{B}_S$, where the index 0 denotes the stationary fields, obeying the equations:

$$\mathrm{Div}\,\widehat{\sigma}(\mathbf{u}_0) = \rho\mathbf{g}, \quad \mathrm{rot}\left(\sigma_e^{-1}(\mathbf{x})\mathrm{rot}\mathbf{H}_0\right) = 0,$$

and the index s denotes the seismic disturbances of the stationary fields.

Then the seismic disturbances $\mathbf{u}_S$ and $\mathbf{H}_S$ obey the equations:

$$\begin{gathered} \mathcal{H}\mathbf{u}_s = 0 \\ \mathbf{B}_{S,t} + \mathrm{rot}\left(\sigma_e^{-1}(\mathbf{x})\mathrm{rot}\mathbf{H}_S\right) = \mathrm{rot}[\mathbf{v} \times (\mathbf{B}_0 + \mathbf{B}_S)] \end{gathered}$$

According to our preliminary calculations (and it is known from observations as well), $\mathbf{B}_S$ is negligible, as compared with $\mathbf{B}_0$, and will be omitted. As we shall consider only the seismic disturbances of the fields (because stationary fields are not of interest here), the index *s* will be omitted as well. Finally, we obtain the 3D string–diffusion system of equations for the seismic disturbances of elastic (string) field **u** and diffusive field of magnetic intensity **H**:

$$\begin{gathered} \rho\mathbf{u}_{,tt} - \mathrm{Div}\widehat{\sigma}(\mathbf{u}) = 0 \\ \mathbf{H}_{,t} + \mathrm{rot}(\alpha(\mathbf{x})\mathrm{rot}\mathbf{H}) = \mathrm{rot}[\mathbf{v} \times \mathbf{H}_0], \\ \alpha(\mathbf{x}) \equiv (\mu_0\sigma_e(\mathbf{x}))^{-1}, \mathbf{v} \equiv \mathbf{u}_{,t} \end{gathered} \tag{2.5}$$

where the left-hand side of the first equation is determined by formula (2.1) and $\mathbf{B}_0 = \mu_0\mathbf{H}_0$ is the known induction of the geomagnetic field. The index s, which we have used to denote seismic disturbances of the stationary fields, is omitted, because we shall consider only the seismic disturbances **u** and **H** of the elastic and magnetic fields, respectively, just beginning from this place.

So, the field Eq. (2.5) of the string–diffusion model of MLPEEA have been formulated.

2.2.3 *Initial and Boundary Value Conditions*

We suppose that the initial conditions, i.e., the ones before the beginning of the seismic events at the moment $t = 0$, are stationary and, therefore, the initial values of the seismic disturbances **u** and **H** are zero. On the upper boundary G_0 of the lithosphere domain R_l, the mechanical load from the atmosphere and the normal derivatives of the magnetic field components are supposed to be zero. On the side (vertical) boundaries of R_l, we consider the zero values for the normal derivatives of all of the seismic disturbances. On the lower boundary, the values of the normal derivatives of the magnetic field components are supposed to be zero, as well as the

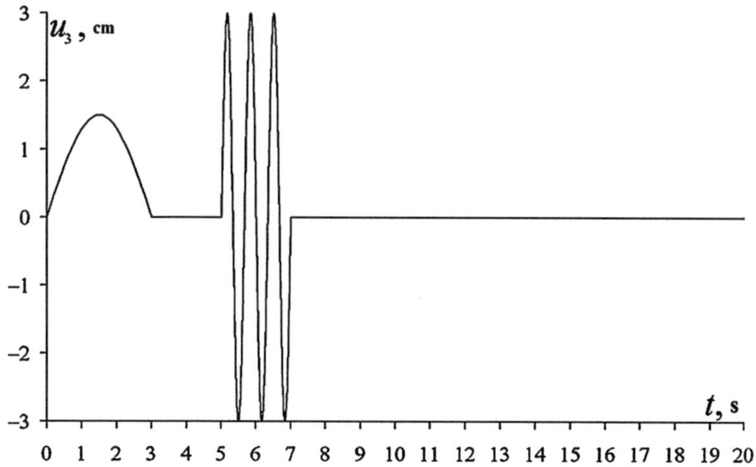

Fig. 2.2 Vertical seismic excitation in the center of the lower boundary of the model lithosphere domain R_1 of the seismo-electromagnetic interaction

values of the horizontal elastic displacements. The only nonzero boundary condition, namely the nonstationary condition of the type of Dirichlet, is determined on the lower boundary $G_m = \{(x, y, z); 0 < x < L_x, 0 < y < L_y, z = L_z\}$ of R_1 for the vertical elastic displacements $u_3(t, x, y, z)$. The energy for the seismo-magnetic interaction is supplied by these displacements:

$$u_3(t, x.y, z)|_{z=L_z} = e^{-b \cdot r^2} \times \begin{cases} A_1 \sin(2\pi f_1 t), & 0 \le t \le 3 \text{ s} \\ 0, & 3 \le t \le 5 \text{ s} \\ A_2 \sin(2\pi f_2 (t-5)), & 5 \le t \le 7 \text{ s} \\ 0, & t > 7 \text{ s} \end{cases} \tag{2.6}$$

These model seismic displacements are represented by two pulses with durations of 3 s and 2 s, amplitudes $A_1 = 1.5$ cm and $A_2 = 3.0$ cm, frequencies $f_1 = 0.1666$ Hz and $f_2 = 1.5$ Hz, $b = 2 \times e^{-9}$, $r^2 = (x - x_c)^2 + (y - y_c)^2$, by the pause between pulses during a period of 2 s (Fig. 2.2).

Because of the exponential multiplier, the amplitudes of the oscillations decrease rapidly with increasing distance from the center $(x_c, y_c) = (80, 80)$, km, of G_m. Therefore, from the sizes of R_1 (Fig. 2.1) and the usual velocities (about 5 and 10 km/s, respectively) of propagation of seismic waves and diffusive magnetic disturbances in our medium (Sect. 2.2.1), a time interval of the order of 10 s is needed to considerably disturb the zero boundary conditions described above, and approximately the same time interval is needed to transfer these boundary disturbances to the domain (let us denote it by B_1, borehole) at the depth of 1 km under the central part of the upper boundary G_0 by $z = 0$. So, the time interval needed to accumulate a considerable change of the magnetic field in B_1 due to action of the disturbed nonzero boundary conditions is more than 20 s. But we "observe" (i.e., calculate) the magnetic field here for 10 s only.

2.2.4 Computed Magnetic Location of a Possible Earthquake Epicenter Area

To get a geometric image of the computed magnetic field in order to estimate coordinates and azimuths (from different points) of the model epicenter area and the seismically activated source of the field, we apply the locating operator for the points (x, y) from the given set M on the plane $z = h$:

$$Pr\,M_{z=h}\nabla B_y(t,x,y,z) \equiv \left\{ \left(\partial B_y(t,x,y,z)/\partial x, \quad \partial B_y(t,x,y,z)/\partial y \right)\Big|_{z=\ h}, \quad (x,y) \in M \right\} \tag{2.7}$$

To construct the geometric image of a magnetic field:

$$\left(B_x(t,x,y,z), B_y(t,x,y,z), B_z(t,x,y,z)\right)$$

at the moment t by mapping the field with this locating operator, it is necessary to calculate gradients of the horizontal component $B_y(t, x, y, z)$ at the moment t for every point $(x, y, z = h)$ by $(x, y) \in M$ and to consider the set of projections of these gradients on the horizontal plane $z = h$. Surely, the ordinate h, the set M of the points with this ordinate, and the magnetic field component B_x or B_y may be chosen in accordance with the case considered. The same is possible for the vertical magnetic field component B_z with simple modifications.

Every projection, i.e., the 2D vector for every fixed point $(x, y) \in M$, may be imaged as the arrow on the plane $z = h$, by a given numerical scale for the components of the projections. Every arrow may be regarded as the local geometric image of the field by mapping the latter with the locating operator $Pr\,M_{z\,=\,h}\nabla B_y(t, x, y, z)$. The set of all local images may be regarded as the global geometric image of the field by mapping the field with this locating operator. In other words, the set of the arrows at points of the set M on the plane $z = h$, constructed in accordance with the right-hand side of the determination (2.7), is the geometric image of the emitted field of magnetic induction determined by the locating operator $Pr\,M_{z\,=\,h}\nabla B_y(t, x, y, z)$ to estimate the coordinates and azimuths of the possible epicenter area and the source of the emitted field.

Evolution of the geometric image in the course of development of a seismic preparation process is of theoretical and practical prognostic interest. The figures below show an example of computed evolution of the geometric image of a seismically generated magnetic field, by the locating operator (2.7).

In Fig. 2.3, at the moment $t = 5.04$ s after the beginning of the model seismic excitation (Sect. 2.3), the maximal value of $B_y(t, x, y, z)|_{z\,=\,h}$, $h = 800$ m, is about 200 pT and the arrows are so small that only the points of the plane $z = 800$ m, where the model borehole magnetometers are disposed, can be seen as a whole. By more detailed analysis, the rather small (about 1 fT/m) arrows, i.e., local images of the

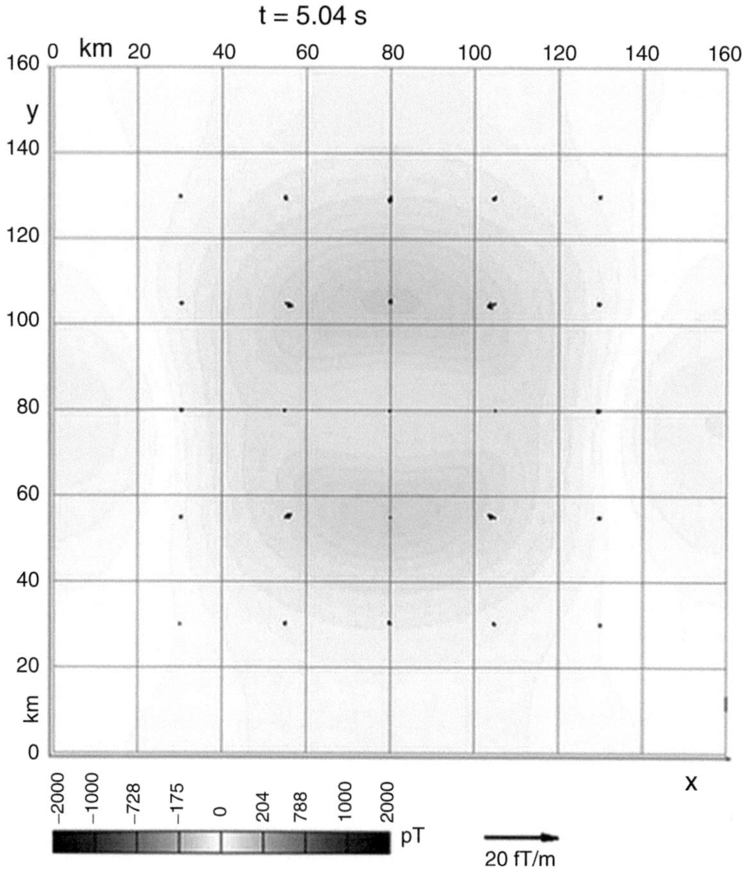

Fig. 2.3 Initial stage of computed magnetic location of the epicenter area, $t = 5.04$ s after the beginning of the model seismic excitation (2.6); see Fig. 2.2. The values of the horizontal component of magnetic induction $B_y(t, x, y, z)$, $z = 800$ m, are imaged with a *gray color*; the geometric image of the magnetic field was constructed with the locating operator (2.7)

field, can be seen at a few points. The gray color scale, applied to show the spatial structure of the field of the horizontal magnetic component $B_y(t, x, y, z)|_{z = h}$, $h = 800$ m, $t = 5.04$ s, and the scale for the field arrows (fT $= 10^{-15}$ T) are shown near the lower boundary of Fig. 2.3. The same scales are applied below.

In Fig. 2.4, by $t = 6.18$ s, i.e., 1.14 s after the picture of a weak field and absence of location in Fig. 2.3, the magnetic field component B_y is strengthened up to 700 pT on the plane $z = h$, because of diffusive propagation of the EM disturbance generated by the seismo-EM transformer E (Fig. 2.1), supplied with energy of seismic excitation (2.6) (Fig. 2.2), in the presence of the geomagnetic field. As a result, the geometric image of the seismic magnetic field, obtained by mapping the field with the locating operator (2.7), demonstrates the distinctive estimation of the coordinates and azimuths (from the different points of disposition of the model magnetometers)

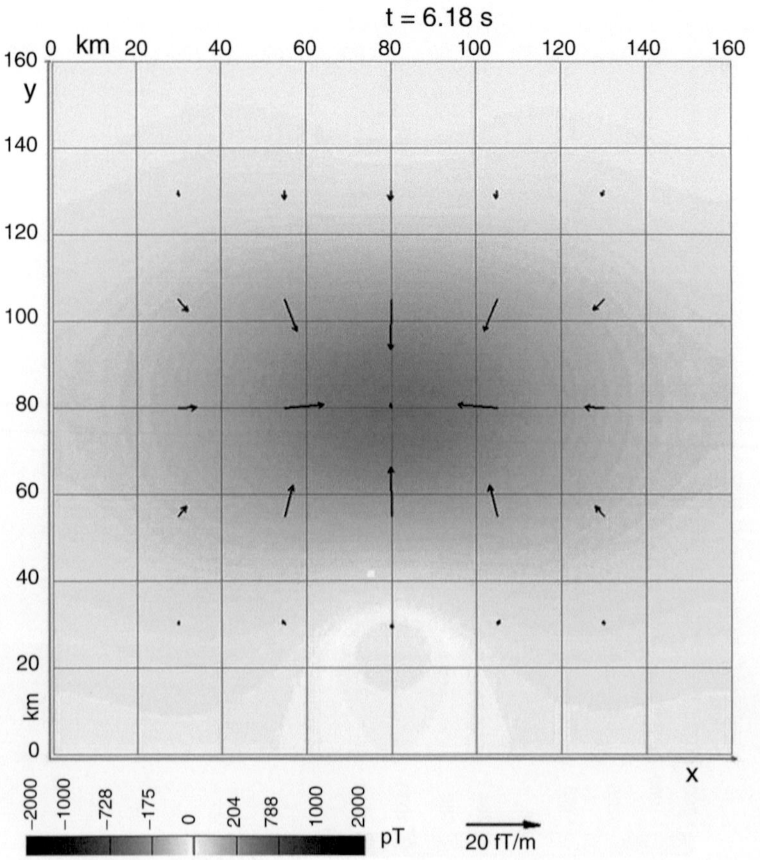

Fig. 2.4 Magnetic location of the model epicenter area, $t = 6.18$ s; other details are explained in the caption of Fig. 2.3

of the area around the center $(x, y) = (80, 80)$, km, of the horizontal cross-section of the model domain R_l. This area may be interpreted as a model of a possible earthquake epicenter area. Let us remember that the center of the seismo-EM transformer E coincides with the center of R_l (see Fig. 2.1). The geometric image of the field is not circular symmetric like the geometry of the domain R_l, including the seismo-EM transformer E (see Sect. 2.1). One can see from Figs. 2.1 and 2.4, taking account of the scales shown there, that the largest arrows (up to 15 fT/m) determine the central, most intensive part of the EM emission from the seismo-EM transformer E.

In Fig. 2.5, at the moment $t = 8.94$ s, one can see the last informative (for the prognosis) stage of the computed evolution of the geometric image. The maximal value of B_y is about 500 pT and the length of the typical arrow corresponds to the value of about 5 fT/m (see the scales); the narrow domain of the maximal EM emission above the seismo-EM transformer has disappeared. Now, the emission is

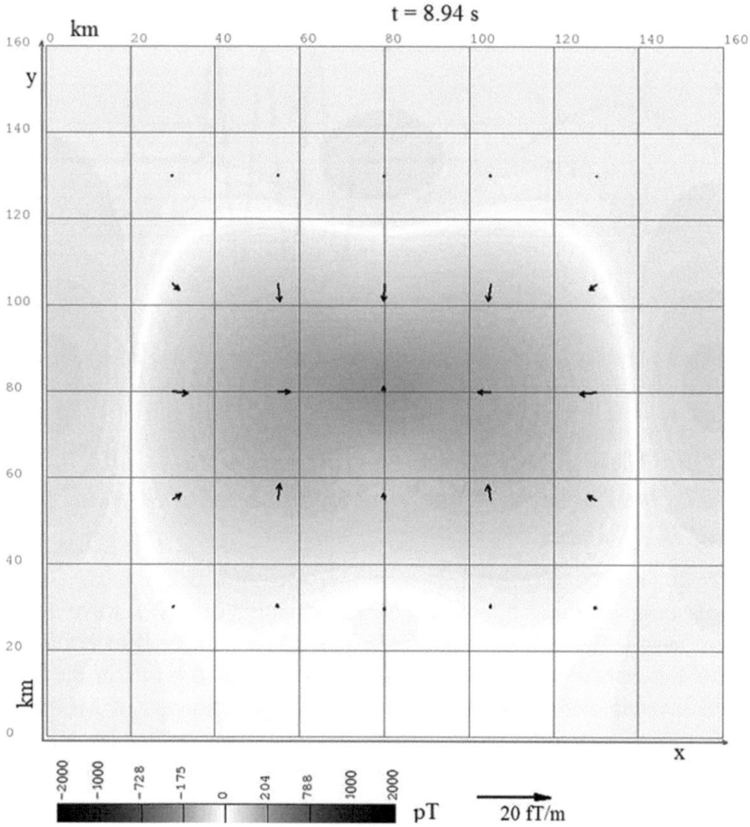

Fig. 2.5 End of the process of magnetic location of the model epicenter area, $t = 8.94$ s; other details are explained in the caption of Fig. 2.3

spatially distributed on the area inside the white curve of zero values of B_y. Outside this curve, the magnetic field changes the sign, but the field is too weak for prognostic analysis. The configuration of the geometric image of the magnetic field under consideration (i.e., the set of arrows) and the zero field curve show the nonsymmetric structure of the model seismically active medium, though the seismic excitation (2.6) is circular symmetric.

Possibly, it is reasonable to consider the combined geometric image of the field including the set of arrows (above) and the curve of zero values of the field's component.

At the moment $t = 10.38$ s (to save the space, we omit the figure), there are arrows with an opposite direction in regard to the computed epicenter area and, therefore, the duration of the informative prognostic evolution of the geometric image is about 3 s with the main conditions considered, namely the lithosphere domain (see Fig. 2.1 and Table 2.1), its seismic excitation (see Fig. 2.2), and the locating operator (2.7).

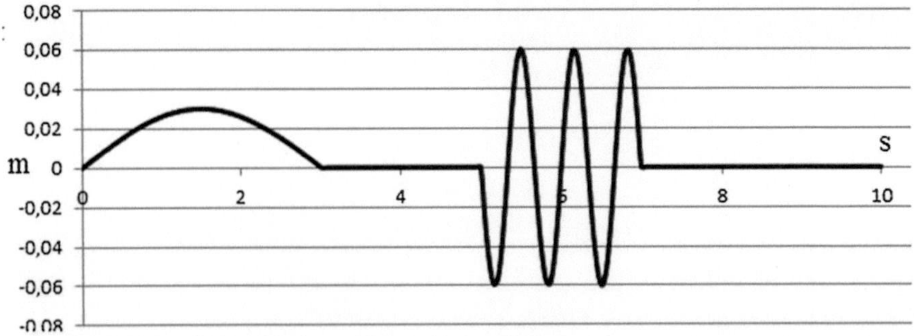

Fig. 2.6 Seismic excitation at the bottom (at a depth of 40 km) of the lithosphere domain by computation of the effect of seismic modulation of the electromagnetic wave

2.2.5 *One More Effect of Seismo-EM Interaction: The Electromagnetic Wave Inherits the Spatial Structure of the Seismic Wave*

We shall describe here the computed spatial modulation of an EM wave in the hertz range by a seismic wave as a result of a seismo-EM interaction process in a model conductive lithosphere domain. This effect was computed by us in the course of theoretical investigation and computations of the phenomenon of MLPEEA (see above). So, the mathematical model of Sect. 2.2.2 has been applied for simulation of the effect of spatial modulation of the EM wave by a seismic one, i.e., the model includes the equations of the 3D dynamic theories of elasticity and Maxwell's electrodynamics with diffusive approximation of the magnetic field.

In our simulations, a weak (precursory) seismic excitation acts, for a few seconds (Fig. 2.6), upon the bottom of a lithosphere domain containing a low-resistivity block (Fig. 2.1). In the presence of the stationary geomagnetic field, the hertz-range EM field is generated mainly by elastic deformations of this conductive block (i.e., seismo-EM transformer).

Spatial modulation of a seismically generated diffusive EM wave by the seismic wave may be regarded as inheritance of the seismic wave's structure by a seismically generated EM wave (Fig. 2.7).

From the mathematical point of view, this effect is caused by the structure of the right side's term of the second equation of system (2.5) including the field of velocity of elastic displacements.

As far as we know, a similar result has not been described in the theoretical works devoted to seismo-electromagnetism (possibly because of the mathematical and computational difficulties of consideration of the 3D dynamical string–diffusion system of PDEs (see Novik (1969) and other references) with the hyperbolic operator of the elasticity theory and the parabolic operator of the Maxwell's electrodynamics without displacement currents). The result of seismic modulation of the EM wave was obtained by us by simulations with a 300-km horizontal size of the

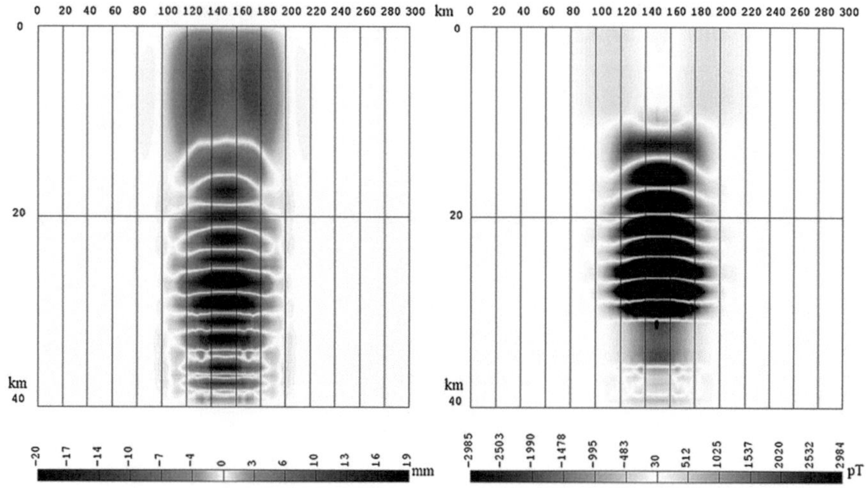

Fig. 2.7 Imprint (i.e., trace) of the seismic wave on an electromagnetic (*EM*) wave generated by a seismic process in the presence of the geomagnetic field. Vertical cross-sections through the center of the domain of the seismo-EM interaction (see Fig. 2.1) are shown. The horizontal size of the domain is 300 km and the depth is 40 km. *Left panel*: vertical displacements in the seismic wave; the *gray scale* is shown in mm. *Right panel*: spatial structure of the horizontal component of magnetic induction B_y of the EM wave inherited from the seismic wave; the *scale* is shown in pT. The time (t) = 9.6 s after the three-dimensional model seismic process starts

lithosphere domain instead of 160 km (Fig. 2.1) and the seismic excitation in Fig. 2.6, which is slightly different from (2.6). Let us note that to check the stability of the seismic modulation of the EM wave, we have applied not only changes in the input data of the initial boundary value problem of the system of PDEs (2.5) but also a change in the dimension of the space of the seismo-EM interaction; namely, the distinctive seismic trace on the seismically originated EM wave was obtained by us in 2D simulations (Novik and Ershov 2001); see Figs 2.8 and 2.9. In this book and other works of the authors may be found details of the consecutive stages of the process of transformation of a weak (precursory) seismic excitation at the upper mantle depth into magnetic signals in the atmosphere.

2.2.6 Discussion about Locating Operators

The variety of the seismic lithosphere structures and the mechanisms of their weak (precursory) seismic excitations (e.g., a non-central P-wave shock in a seismo-EM transformer) is too large to try to describe here the class of cases in which MLPEEA will be effective, as well as the class in which it will not be effective. Surely, the class of the expected effective location will be wider and the location will be more definite if a few locating operators are applied simultaneously—say, horizontal projections

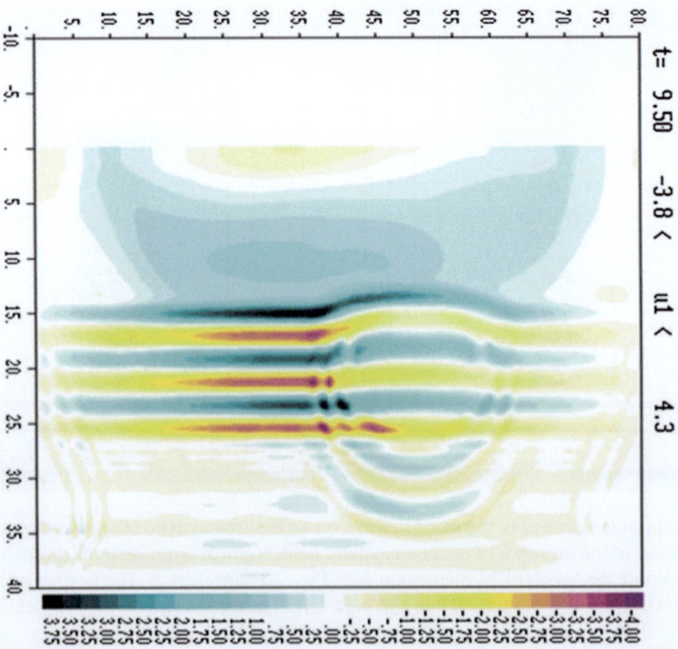

Fig. 2.8 Two-dimensional model seismic wave of vertical displacements, in the range of −3.8 to 4.3 cm, from when the first aperiodic pulse (at a depth of 40 km) of the seismic excitation (see Fig. 2.2) reaches the earth's surface (at $t = 9.50$ s), is reflected from it, and then shocks the upper part of the conductive block (a typical structure of seismically active lithosphere zones). At the same time, the seismic wave from the periodic second pulse deforms the block. Both of these interactions occur in the presence of the geomagnetic field

of gradients of all magnetic components along with their lines of zero values. Maybe the horizontal projections of the perpendiculars to the "front" of a given magnetic component (or a group of them) will be more effective, with some conditions in regard to the medium and seismic excitation, than the locating operator determined by the gradient [e.g., formula (2.7)]. Let us remember that a real wave front is absent in our case of diffusive propagation of the EM field in the lithosphere. The surface from one side of which (taking account of the direction of propagation) the field values are too small to be measured by the devices applied may be regarded as the "front" of a diffusive EM disturbance. The surface of a fixed value of a diffusive field (or a few similar surfaces) may be applied to obtain an informative set of perpendiculars.

The important and difficult problem of the optimal (including the prognostic value and universality) locating operator or locating process, with an adaptive group of operators, is not within the scope of this work. Other problems of earthquake prognosis and approaches may be found in Hayakawa (2015).

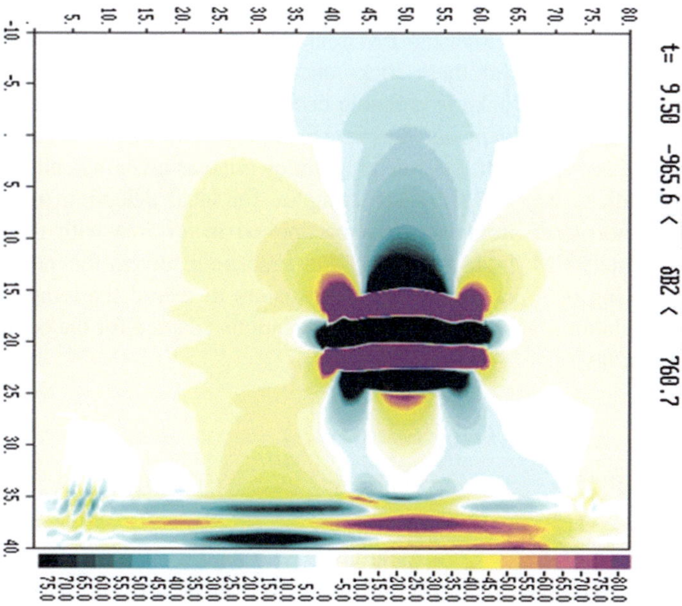

Fig. 2.9 Magnetic induction emission from the seismically disturbed conductive block (see the capture of Fig. 2.8). The depth in the lithosphere and the height in the atmosphere (with −) are in km, and the earth's surface level is 0; at $t = 9.50$ s after the beginning of the seismic excitation (see Fig. 2.2), the range of the horizontal component of the seismically generated magnetic induction is −956.6 to 760.7 pT

Possible Applications of the Results Let us note that the numerical values of the computed horizontal projections of the low-frequency magnetic field gradients (1–15 fT/m for the different stages of locating and different disposition of magnetometers; see Figs. 2.3, 2.4, and 2.5) correspond to the values observed, i.e., 1–5 pT/km (Ismaguilov et al. 2001, 2003). We recognize that the characteristics of the medium, seismic excitation, and other conditions were not (and may not be) known well by the field experiments (Kopytenko et al. 1999, 2000). We considered here the simplest characteristics (i.e., the input data of the mathematical model and computations) acceptable from the geological, geophysical, and petrophysical viewpoints.

We believe that if some area of a seismically active region is recognized by an available seismo-magnetic locating system (a network of locating magnetic stations and seismographs) as an area with repeated low-frequency EM emissions, then this area may be the epicenter area of a future earthquake. Indeed, these emissions are generated by local decreases in mechanical stress in the vicinity of a seismo-EM transformer, which is being deformed by these seismic excitations in the presence of the geomagnetic field. So, repeated EM emissions, especially with specific seismic modulation (see Sect. 2.2.4), are to be interpreted as the sign of accumulation of additional deformations not too far from the center of the emitting area, because the seismic preparation zone as a whole is under stress and propagation of local excitations over long distances is hardly possible. The process of accumulation of deformations in rocks near a seismo-EM transformer may create a deformation value above the critical level for these rocks and, as a result, an earthquake. Let us emphasize that an earthquake is probable because the repeated seismic excitations in a given lithosphere domain, and the corresponding accumulation of deformations, result from the activity of the seismic preparation zone as a whole, and this activity may be periodical. So, a multisensor complex for local detection of precursory signals and short-term prognosis must be applied in an area with repeated EM emissions, namely EM (high-frequency) and acoustic receivers for monitoring the possible cracking of rocks, quartz thermo-resistors to record the temperature and heat flow oscillations, and sounding instrumentations adapted for the continental or marine conditions.

2.3 Conclusion

1. On the basis of the known geographic correlation between high-seismicity lithosphere zones and ones with low-resistivity structures (electrical conductivity of the order of 0.1 S/m with typical size and depth of the order of 10 km) and applying the magneto-thermo-elasticity theory, we have formulated a relatively simple mathematical model of magnetic location of a possible earthquake epicenter area (MLPEEA).
2. The calculations with this model demonstrate the observed (see Kopytenko et al. (1999, 2000) and other references) seismo-electromagnetic phenomenon of MLPEEA.
3. The electromagnetic wave inherits the spatial structure of the seismic wave during seismo-electromagnetic interaction.
4. The effect of MLPEEA and the effect of seismic modulation of the electromagnetic wave are stable against uncertainty of the geometric and physical input data of the model.

References

Arora, B. R., & Mahashabde, M. V. (1987). A transvers conductive structure in the northwest Himalaya. *Physics of the Earth Planetary Interior, 45*, 119–127.

Hayakawa, M. (2015). *Earthquake prediction with radio techniques*. Hoboken: Wiley.

Ismaguilov, V. S., Kopytenko, Y. A., Hattori, K., Voronov, P. M., Molchanov, O. A., & Hayakawa, M. (2001). ULF magnetic emissions connected with under sea bottom earthquakes. *Natural Hazards and Earth System Sciences, 1*, 23–31.

Ismaguilov, V. S., Kopytenko, Y. A., Hattori, K., & Hayakawa, M. (2003). Variation of phase velocities and gradient values of ULF geomagnetic disturbances connected with the Izu strong earthquakes. *Natural Hazards and Earth System Sciences, 3*, 211–215.

Kopytenko, Yu. A., Ismaguilov, V. S., Voronov, P. M., Kopytenko, E. A., Molchanov, O. A., Hayakawa, M., et al. (1999). Magnetic disturbances in ULF range connected with seismic sources. In *Proceedings of the 2nd International Conference on Marine Electromagnetics—Marelec*. 99, France (pp. 435–445).

Kopytenko, Y. A., Ismaguilov, V. S., Kopytenko, E. A., Voronov, P. M., & Zaitsev, D. B. (2000). Magnetic location of geomagnetic disturbance sources. *Geophysics, 371*(5), 685–687.

Maugin, G. A. (1988). *Continuum mechanics of electromagnetic solids*. Amsterdam: North-Holland.

Novik, O. B. (1969). The Cauchy problem for a system of PDEs including the hyperbolic and parabolic operators. *Journal of Computational Mathematics and Mathematical Physics, 1*, 122–136. (In Russian. Задача Коши для системы УРЧП, содержащей гиперболический и параболический операторы. ЖВМиМФ, 1969, No 1, с. 122–136).

Novik, O. B. (1994). *Seismogenic magnetothermoelastic effect*. In Doklady (Trans) Russian Academy of Sciences, 338(2), 238–241 (In Russian).

Novik, O. B. (1995). *The Galerkin method for a three-dimensional nonlinear system of magnetothermoelasticity equations*. In Doklady (Trans) Russian Academy of Sciences, 335A, 15–21 (Translated from Russian: Doklady Rossiyskoy Akademii Nauk, 1994, 334, No 2, pp. 100–102).

Novik, O. B., & Ershov, S. V. (2001). *Electromagnetic and temperature signals from the Earth's depth* (255 p.). Moscow: Scientific Monographs of Federal Program of Integration of High Education and Fundamental Science (In Russian).

Robb, L. (2005). *Introduction to ore-forming processes* (368 p). Hoboken: Blackwell.

Schön, J. (2011). Physical properties of rocks: A workbook. In *Handbook of petroleum exploration and production* (Vol. 8, 481 p.). Amsterdam: Elsevier.

Chapter 3
Tsunami Precursors: Mathematical Model and Applications

Abstract A mathematical model of seismo-hydro-electromagnetic (EM)-temperature geophysical field interaction has been formulated, according to the principles of the magneto-thermo-elasticity and magneto-hydrodynamic theories. Signals of different physical natures, arising in a lithosphere–ocean–atmosphere domain as a result of submarine seismic excitation, were computed. On the basis of satisfactory results of comparisons with measurements, a multidisciplinary vertically distributed (from the seafloor up to the ionosphere) tsunami precursors detection system is described.

Keywords Tsunami · Precursors · Detection · Mathematical model

3.1 Introduction

Most of the earth's population is located in near-sea regions, and this proportion will increase permanently during the coming decades, according to the majority of estimations. Hundreds of thousands of fatalities and distructions (including that of the Fukushima I nuclear power station during operation in 2011), caused by seismogenic releases of huge energy from deformations of the ocean lithosphere plates (these phenomena, with a seismic source under the seafloor, are known as *underwater earthquakes* or *seaquakes*) demonstrate that the prognosis of similar dangerous geodynamic events is far from satisfactory, though new monitoring systems have been applied, including one based on the Global Positioning System (GPS), since the Indian Ocean tsunami in 2004 (which caused about 225,000 fatalities). A typical size of a strong earthquake preparation zone is of the order of 1000 km, and we suppose that the Haiti earthquake in 2010 (which caused about 250,000 fatalities), with an epicenter located on land at a distance of 22 km from the shore, was supplied by energy from deformations of the ocean lithosphere, as well as seaquakes, as mentioned above.

Effective detection of precursors of dangerous stages of accumulation of deformations is not possible without understanding the physics of this process; namely, a

O. Novik et al., *Electromagnetic Geophysical Fields*,
https://doi.org/10.1007/978-3-319-98461-2_3

seismogenic deformation process under the ocean bottom may not be regarded as a purely mechanical one. Indeed, the medium under the seafloor is not static; it is moving more or less in different parts and during different time intervals, in the course of accumulation of seismogenic deformations. The motion of the sea bottom geological structures is being transferred into the seawater (with conductivity from 3 to 4 S/m) and an electromagnetic (EM) field is arising there, in the presence of the geomagnetic field, as a component of the seismogenic deformation process. Moreover, the rocks of the ocean geological structures are not ideal materials from the rheological viewpoint and their stress–strain relations may include nonmechanical terms. In an attempt to understand the seismo-hydro-electromagnetic and temperature (T) interaction (SHEMTI) process accompanying ocean lithosphere seismogenic deformations, we have formulated a mathematical model—to be denoted as the SHEMTI model—of interaction of geophysical fields of different natures in the lithosphere–ocean–atmosphere medium (see Appendices A and B). Time series of precursory SHEMTI model signals generated by weak seismic excitation (disturbance of a stationary configuration of the lithosphere deformation field) in the oceanic upper mantle of a model lithosphere–ocean–atmosphere domain are shown in Sect. 3.2. A scheme for a tsunami precursors detection system, corresponding to the characteristics of computed signals (the latter, in turn, correspond to observations) is described in Sect. 3.3.

We use the term *tsunami precursors* in the following sense. First of all, a tsunami precursor is a precursor of a marine earthquake (i.e., seaquake), and a precursor of an earthquake is understood as a measurable signal of geological and/or geophysical origin in the area of preparation of an earthquake that occurs before the earthquake, regardless of the prognostic value of the signal under consideration. Certainly, this is not a very precise definition, according to the level of the established prediction science, with its prognostic failures. Though most tsunamis are caused by seaquakes, or by landslides and volcanic eruptions triggered by seaquakes, the latter are not always accompanied by a considerable tsunami. The problem of classification of seaquakes (not only extremely strong ones) and their precursors in regard to the potential for tsunami generation has not been resolved, because of different underwater seismic source particularities, the seafloor relief, and so on. In any case, we should consider seaquake precursors as ones of tsunamis. Besides, seaquakes are destructive even without considerable tsunamis. With regard to tsunami data and forecasting, see http://www.prh.noaa.gov/itic/tsunami_events/events.html.

This chapter should not be regarded as a project on a tsunami precursors detection system. It includes only some physical and mathematical arguments to formulate a technical task for a similar project.

As far as we know, a description of a multidisciplinary vertically distributed (from the seafloor up to the ionosphere) tsunami precursors detection system, based on a mathematical model of seismo-hydro-EM-temperature interaction that corresponds to observations, is new.

3.2 Computed Signals of the Model Seismic Submarine Excitation

The time series of the seismic (i.e., seismically originated) signals at the seafloor shown in this section have been computed on the basis of the SHEMTI model and will be used in the next section for description of the scheme of the tsunami precursors detection, because the computed seismic, EM, temperature, and hydrodynamic signal characteristics correspond to observations (Belov et al. 1974; Gohberg et al. 1991; Hayakawa 2003; Iyemori et al. 1996). These signals have been computed on the basis of a rather weak seismic excitation represented by elastic displacements, with amplitude and duration of the order of a few centimeters and seconds, respectively, in the upper mantle under the sea (Appendix B). We believe a similar excitation may be generated by different dynamical processes under the ocean bottom, in particular in the course of accumulation and redistribution (e.g., local releases) of deformation energy of the nonuniformly strained media of interacting ocean lithosphere plates. The computed seismic EM signal amplitude is increased by stronger seismic excitation approximately proportionally to its amplitude. But weak (precursory) signals are considered here to estimate the sensitivity thresholds of the appropriate receivers.

The question about the first measurable seismic signal, i.e., the first measurable precursor of a tsunami (see Sect. 3.1), is of the utmost importance for construction of an early detection system. Indeed, what are the characteristics of the first precursors, to determine the appropriate receivers, and where they should be located? According to our computations, the first precursors are the low-frequency EM signals within the frequency range of the seismic submarine excitation (from 0.1 to 10 Hz in the case discussed in Appendices A and B), but it is not true that the seafloor is the only place where the first precursors may be registered, though these signals are propagated from underneath it.

To answer these questions, let us briefly consider a few details of the computed SHEMTI process. According to our mathematical and numerical investigations of the initial boundary value problem for the system of PDEs of the SHEMTI model (Appendix A), the characteristics of this process are stable against reasonable variations (including the physics of rocks) of the mechanical, EM, and thermophysical data on the lithosphere–ocean–atmosphere medium.

After starting from the submarine seismic source and outrunning the seismic wave, the EM signal arrives at the seafloor, namely at the top of the sedimentary layer. This signal is "frozen" there. Its front replicates the configuration of the sea bottom surface, without any considerable penetration into the seawater, because of its high electrical conductivity, during the time interval of about 1.5 s needed (according to the conditions of Appendix B) for the delayed seismic P wave of vertical elastic displacements to arrive at the seafloor as well. The energy supplied by this arrival (in other words, by the seismic P wave shock in the seafloor; see Fig. 3.1) is sufficient to elicit EM oscillations at the top of the sedimentary layer (Fig. 3.2) and

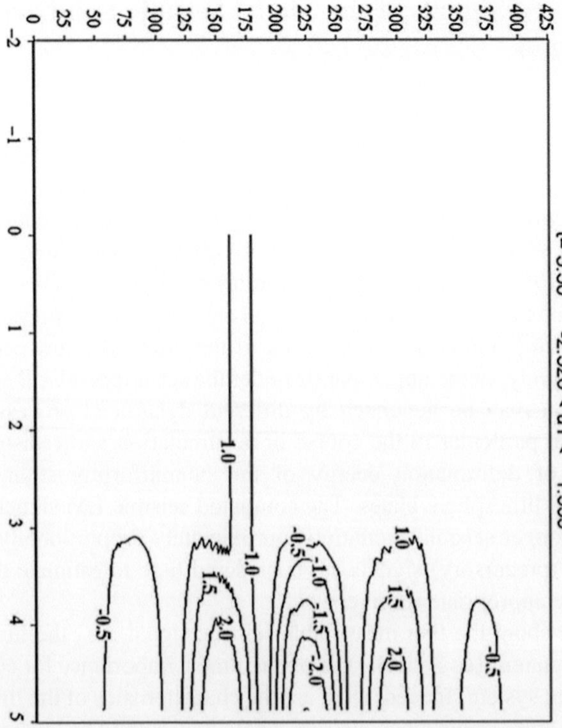

Fig. 3.1 Contour plot of the field u_1 (in cm) of vertical displacements in seawater at the moment $t = 5.30$ s after a submarine earthquake begins: a narrow domain of the vertical upward seawater motion arises above the initial contact, at $t = 5.25$ s, of the front of the seismic wave of vertical displacements on the seafloor. Induction of the geomagnetic field in this moving domain of seawater is the main mechanism, according to the seismo-hydro-electromagnetic-thermal interaction (*SHEMTI*) model, of the seismo-electromagnetic emission from the sea surface. Horizontal coordinates are shown at the upper boundary in km (see Fig. 3.4); vertical coordinates (in km) are positive under the sea level (marked by 0) and negative above it

EM emissions (Fig. 3.3) from the sea surface, as a result of the SHEMTI process in seawater in the presence of the geomagnetic and gravity fields.

Figure 3.2 demonstrates the first magnetic precursors (with an amplitude of the order of a few hundred pT), which may be measured, according to our computations, at a seafloor point with an epicenter distance of 12.5 km, i.e., the distance of the point under consideration from the seismic epicenter at the seafloor. In turn, this epicenter is the vertical projection of the center of the seismic source (the point of the maximum amplitude of the seismogenic displacements, in our model) at the seafloor. But a similar case, when a seabed station (or a network of stations) is located near the epicenter at the seafloor, is not the most typical.

Figure 3.2 shows that the seismic magnetic signal at the sea bottom increases considerably for 10 s after the beginning of the seismic activation of the upper mantle under the sea bottom, e.g., up to 240 pT for the vertical component B_1 of

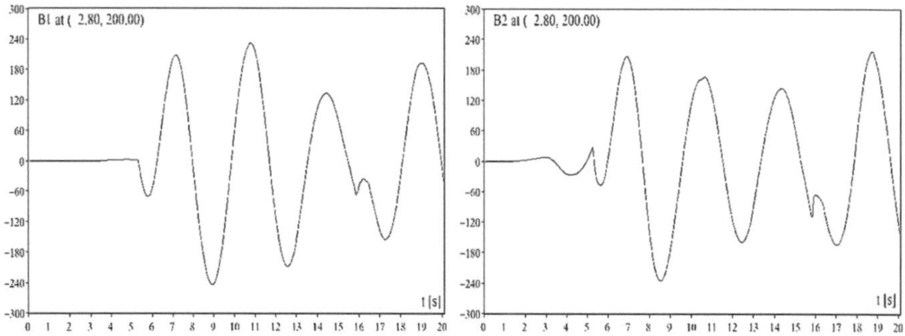

Fig. 3.2 Time series of the magnetic signal components B_1 (vertical) and B_2 (horizontal) (in pT) at the sea bottom point at a depth 2.80 km from the sea level and at a distance of 200 km from the western shore (12.5 km from the epicenter). The measurable (about 50 pT) oscillations of the horizontal component B_2 (in pT) of magnetic induction are detected 4 s after the beginning of the seismic excitation at $t = 0$ (see Appendix B). The beginning of the oscillations of the vertical component B_1 is delayed with respect to B_2

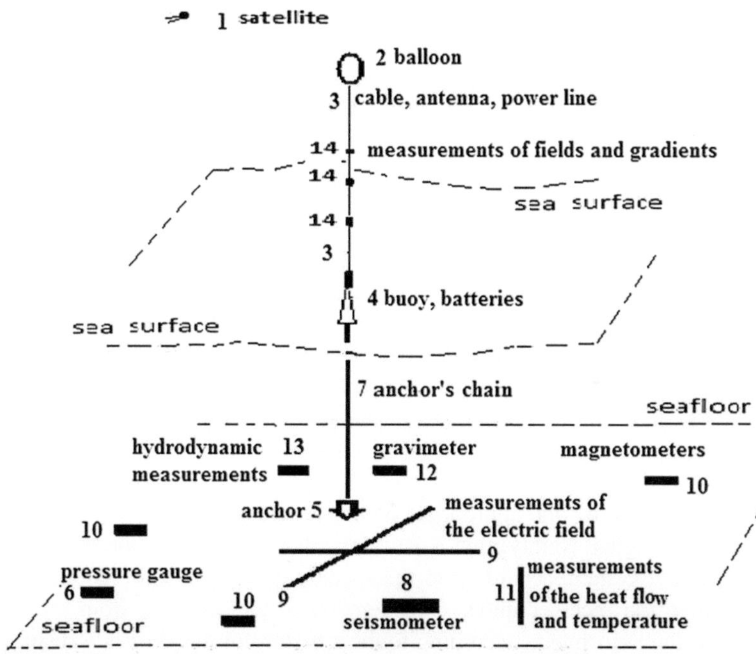

Fig. 3.3 Block scheme of the tsunami precursors detection system (see Sects. 3.3.2, 3.3.3, and 3.3.4). *1* satellite complex, *2* moored balloon, *3* the top part of the cable and devices on it, *4* anchored buoy, *5* anchor, *6* tsunameter with a water pressure gage, *7* anchor chain, *8* seismometers, *9* electromagnetic block with electrical field sensors, *10* magnetometers, *11* thermometry block with quartz thermo-resistors, *12* gravimeter, *13* hydrodynamic block with an acoustic Doppler velocimeter

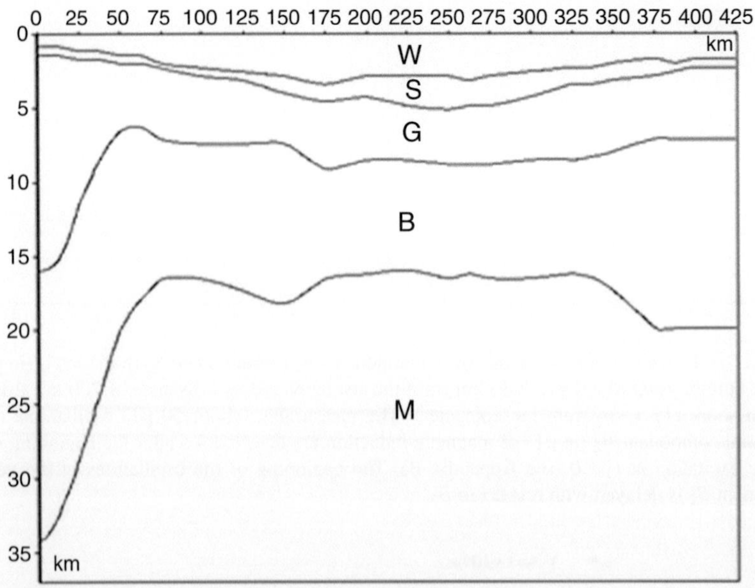

Fig. 3.4 Seawater (*W*) and geological structure of the lithosphere part; the horizontal coordinates are in km. The cross-section end points are 43.2° N, 132° E and 39.7° N, 133.7° E

magnetic induction at the point with a horizontal coordinate $x_2 = 200$ km, $x_2 = 0$, at the western shore of the sea (Fig. 3.4). It is reasonably to note that seismic EM signals at the seafloor are the only signals of that kind that (unlike signals at the sea surface and above) are protected by seawater from ionospheric EM disturbances and from the EM noise of a hard sea.

The oscillations of the vertical and horizontal seismic magnetic components at the sea surface at a point with a distance of 12.5 km from the epicenter, located at the sea surface, are similar to their oscillations at the sea bottom (Fig. 3.2), regarding the amplitudes and main frequencies. Possibly, the EM noise—due to wind waves or other phenomena—will distort the seismic EM signals at the sea surface (recorded by a receiver on a buoy) and appropriate hardware and software must be applied to extract a useful signal. Surely, this is true for other receivers in the tsunami precursors detection system, according to the specific dynamics of the domain of the medium where the receiver under consideration is located, i.e., from the ocean bottom up to the ionosphere, according to the structure of the vertically distributed detection system (see Sect. 3.3).

At a height of 10 km in the atmosphere above a nearly central point of the area of seawater concerned, the computed vertical component of the disturbance of magnetic induction B_1 is up to 160 pT at the moment $t = 11$ sec and the horizontal component B_2 is up to 80 pT at this point at the moment $t = 9$ s.

In contrast to the examples above, let us now consider the seismic magnetic signal at the point above the sea (at a height of 10 km) with a considerable distance of

112.5 km from the epicenter on the sea surface—this epicenter being the vertical projection of the seafloor epicenter.

Naturally, this seismic magnetic signal arises practically at the same time as signals in the epicenter domain because of the light velocity of EM signal propagation in the atmosphere. From the mathematical point of view, in the SHEMTI model, EM signals in the atmosphere are described by the Laplace equation (conjugation conditions with the nonstationary EM field of the lithosphere are used), which belongs to the elliptic type of PDEs characterized by an infinite velocity of propagation of disturbances. But in the lithosphere without insulators (our assumption about the considered medium), low-frequency EM signals are described by the Maxwell equations of magnetic diffusion (i.e., without displacement currents) of the parabolic type with a rather slow velocity of propagation of measurable signals. So, a seafloor receiver, with the same epicenter distance (of about 100 km) as that of the receiver above the sea, will register a seismic magnetic signal later.

The computed magnitude of the amplitude of the temperature oscillations, accompanying the seismic and EM wave propagation in the lithosphere part of the model medium, is of the order of a few millikelvins at the top of the sedimentary layer, whereas their main frequencies are in the frequency range of the seismic excitation, as well as other signals considered according to the conditions described in Appendices A and B. Similar temperature disturbances do not accompany the ionosphere EM signals at the seafloor and should be measured along with the heat flow variations as a sign of an ocean lithosphere seismic excitation.

More detailed physical and mathematical background information on the seismo-hydro-EM-temperature interaction approach to the problem of seaquake and tsunami precursors may be found in various papers (Novik 1995; Novik et al. 2001; Ershov et al. 2001, 2006; Ershov and Novik 2009) and in a previous book by Novik and Ershov (2001).

3.3 TSUNPREDET System

3.3.1 General Characteristic of the System

According to the results of the SHEMTI model (see Sect. 3.3.2 and Appendices A and B), the first signal of a weak seismic submarine excitation may be registered by the receiver in the seafloor complex or by receivers in the complexes located higher, depending on the epicenter distance from the seafloor station and the sea depth. Therefore, a multidisciplinary vertically distributed seaquake/tsunami precursors detection (TSUNPREDET) system must be distributed along a vertical line beginning from the seafloor complex, recording original lithosphere signals in its vicinity only, and up to the balloon-borne and satellite-borne complexes, recording signals from a considerable area, i.e., from remote seismic sources, maybe. But these signals are modified by the electro-conductive seawater and ionosphere media with their own dynamics, which are not connected closely with the lithosphere dynamics, unlike the seafloor signals. In other words, it is necessary to analyze the recordings of the balloon-borne and satellite-borne receivers along with those of the receivers in the seafloor complex to extract the particularities of the recordings above the sea.

So, the TSUNPREDET system (see below) includes seafloor, sea surface, balloon, and satellite complexes. Each of these complexes includes receivers for retranslation of information to a complex above it or to a processing center (the only case for the satellite) and sensors for detection of seaquake precursors. Receivers for retranslation are not discussed below, and the word *receiver* refers to a sensor of possible EM precursory signals. Because of the distribution of the multidisciplinary sensors, i.e., sensors for fields of different physical natures, along the vertical line, the problem of separation of ionosphere EM signals (e.g., caused by ionosphere Pc pulsations, with their frequency range being similar to that of lithosphere EM signals) may be solved by precise measurements of the arrival times of signals at the seafloor, balloon, and satellite. In addition, let us take account of the slow diffusive propagation of low-frequency EM signals in seawater (Novik et al. 2001; Ershov et al. 2001, 2006). Moreover, the multidisciplinary recordings provide a tool to distinguish false EM signals (see Sects. 3.2 and 3.3.2).

Because of the compliance between the SHEMTI model signals and measured ones published by other authors (see Sect. 3.2), our approach to the TSUNPREDET system is based on calculations and measurements. Very-high-frequency (VHF) earthquake precursors (Ruzhin and Nomicos 2007) have been observed and used for prognosis, and VHF receivers are foreseen for the TSUNPEDET, as well as low-frequency ones. Here we show that a similar system may be composed that is largely based on known instrumentation (below) and software. The latter includes, mostly, algorithms for receiving signals of different physical natures, analysis of multidimensional time series, and solution of multidisciplinary (i.e., joint) inverse problems to detect changes in processes and the structure of the medium; also, network algorithms are of importance (Siripunvaraporna et al. 2005; Basseville and Benveniste 1986; Casey et al. 2008).

Details regarding changes of regimes of functioning according to the data received (adaptation to signs of a seaquake preparation) or the stability of the TSUNPREDET system (e.g., different parts of the system with similar functions for replacements) are omitted from this description. Let us note that we use the expression *seafloor complex* as a synonym for the term *seabed station*.

3.3.2 Seafloor Complex

This complex of the TSUNPREDET includes:

- A tsunameter with a water pressure gauge (numbered 6 in Fig. 3.3).
- An anchor (numbered 5 in Fig. 3.3) and an anchor chain (numbered 7 in Fig. 3.3) (see a modification in Sect. 3.3.6).

- A block of seismometers (numbered 8 in Fig. 3.3) with a strong motion seismometer and broadband one, both of them being controlled in regard to their contact with the ground, especially after seismic events.
- An EM block with electrical field sensors (numbered 9 in Fig. 3.3) and magnetometers (numbered 10 in Fig. 3.3).
- A thermometry block, including measurement of variations (of the order of a few millikelvins) in temperature and heat flow with quartz thermo-resistors (numbered 11 in Fig. 3.3) and special pivots to immerse them in sediments; a modification with joint temperature and electrical potential measurements is seen in Sect. 3.3.6 (let us remember that the ionosphere EM signals at the seafloor are not accompanied by temperature disturbances, unlike seismic EM signals, so false EM signals may be separated).
- A gravimeter (numbered 12 in Fig. 3.3), e.g., with a mechanical spring for a proof mass, to measure the vertical component of tides and to record earthquakes and teleseisms.
- A hydrodynamic block with an acoustic Doppler velocimeter (numbered 13 in Fig. 3.3) to deal with changes in the hydrodynamic field near the seafloor and the devices there (Beransoli et al. 2002; González et al. 2005; Landau and Lifshits 1986; Lin et al. 2011; Sherwood 2011; Toh et al. 1998).

Let us comment on this short description. According to the SHEMTI model, the amplitude of the seismic (seismically originated) diffusive magnetic low-frequency (0.1–10 Hz) computed signals at the seafloor and above, measuring a few hundred picoteslas, caused by a weak seismic excitation in the upper mantle under the sea, increases proportionally (approximately) to the excitation amplitude. So, a suitable magnetometer (see above) must be included in the EM block in the seabed station of the detection system under consideration. As a matter of fact, considerable physical and engineering experience has been accumulated in the course of application of magnetometers in seawater. All types of magnetometer technologies (including advanced ones based on the Overhauser effect sensor or on the highest-sensitivity cesium vapor sensor), gradiometer configurations, methods of control and positioning, digital data transmission, and measured data inversion have been applied by marine exploration geophysicists in the case of seabed conditions.

From our viewpoint, a reasonable solution is to include in the seafloor complex a system of magnetometers—say, three of them (numbered 10 in Fig. 3.3). A few magnetometers are needed not only to provide for replacement of a magnetometer that is found to be malfunctioning (perpetual true recording and data transmission under automatic control are important abilities of a detection system) but also to estimate the direction along which the full seismic (seismically originated) magnetic field or its components are increasing. In other words, the epicenter azimuth will be estimated by this approach, according to our preliminary calculations for the problem of an epicenter.

Three-component magnetometers are needed to understand if there is a specific polarization of precursory magnetic signals, whereas more sensitive magnetometers measuring a full field only may be useful to clarify precursory particularities of small field variations at the seafloor, if there are any.

Marine magneto-telluric (MMT) sounding (see Constable et al. 1998) is used as one of the marine exploration geophysical methods to estimate spatial particularities of electrical conductivity under the seafloor, and we hope it can be effective for detection of temporal changes of conductivity caused by seismic deformations. Because of the high conductivity (about 3.5 S/m) of seawater, short periods and detailed survey may not be applied. But it is not a strong limitation in the case under consideration because large-scale, deeply immersed ocean lithosphere structure information, available from long-period MMT sounding, are of main interest in predicting oceanic geodynamic events.

On the other hand, for the same reason, the EM field sensors on a deepwater seafloor are not influenced considerably by EM signals caused by *short* hydrodynamic waves (a hard sea). Meanwhile, we believe that seabed EM sensors can detect tsunamis, which are *long* hydrodynamic waves. Indeed, seabed pressure sensors [e.g., a quartz crystal vibrator on a tsunameter, which is used in the widespread Deep-Ocean Assessment and Reporting of Tsunamis (DART)–type detection systems (González et al. 2005)] do register pressure changes of the order of 0.01 m at a full sea depth up to 7000 m.

Therefore, a tsunami wave which, far from a shore, is a long (e.g., 100,000 m) hydrodynamic surface wave of a small (e.g., 0.5 m) amplitude, with a period of the order of 1000 s, may be registered by a seabed pressure sensor when the wave is propagating above it. To be more accurate, let us note that the mechanical effects of a surface wave, as mentioned, will appear at the seafloor with a delay of a few seconds needed for the hydro-acoustic signal of a tsunami wave to propagate along the 7000-m-long trace in normal salinity. Possibly, the EM induction effect of the arrival of a tsunami hydro-acoustic signal at the seafloor may be registered by the EM sensors. This possibility can be clarified by the mechanical equipment (the pressure sensor and the seismographs) and the EM equipment (the magnetometers and the electrical field sensor) in a seafloor complex. It must be remembered that registration of a developed stage of a seaquake–tsunami process in the form of a tsunami wave may be regarded as one of the goals of the considered system. In the case of a short epicenter distance from the seabed station, the main goal is to register seismically originated signals of different physical natures coming from under the seafloor.

The known diffusive character of EM field propagation in the conductive tectonically active ocean lithosphere medium (we have observed this diffusivity in our calculations as well) does not allow us to obtain a spatial image of the investigated medium with the accuracy of seismic instrumentation. But, given that the goal is to detect an undersea earthquake preparation process, we are interested mostly in recording temporal changes of the MMT sounding results due to the sub-seabed dynamics.

These measurements of temporal changes do not suffer that much from the strong distorting influence of small particularities of the seafloor relief, because of the large difference in the seawater and sedimentary conductivities. Also, other difficulties with MMT and other exploration methods are not that crucial in our applications, e.g., industrial EM noises (a typical difficulty in land magneto-telluric soundings) are suppressed at the seafloor.

Ocean lithosphere sounding by controlled source of EM (CSEM) fields would be informative in regard to dynamical processes under the sea bottom, but for long-term unattended observations at the seafloor, there are considerable difficulties with electrical power sources. A suitable one for CSEM soundings may be provided by a ship, e.g., during replacement of the batteries. The batteries for the seafloor, sea surface, and balloon complexes may be located on a buoy connected by power transmission lines to the remote complexes mentioned, because the solar panels on the buoy and balloon may not provide sufficient power. Let us note that the power and measured data transmission may be combined in the same cable. Transmission via fiber-optic cabling is effective in seafloor conditions and may be used for connection of the seafloor devices with the buoy, but such transmissions are omitted from Fig. 3.3. Also, a tiltmeter is not included in the tsunami detection scheme shown in Fig. 3.3.

Data from a magnetometer immersed in a submarine borehole, i.e., nearer to a possible seismic source, will be received earlier and without attenuation of the signal by propagation along the upper part of the lithosphere trace to the top of a sedimentary layer (with its higher electrical conductivity because of penetration of seawater), where the EM block of the seafloor complex is located (Hayman et al. 2010).

3.3.3 Sea Surface Complex

This complex includes an anchored buoy (numbered 4 in Fig. 3.3) with batteries and solar panels for the seabed and balloon instrumentation (see above). In addition, the sea surface complex contains the junction of the underwater cable with that of the balloon to transmit the data from the seabed station to another balloon, a satellite, or a data-processing center onshore, depending on the distances. These communications are bidirectional to control the proper functioning of all of the sensors, receivers, transmitters, and energy sources. The buoy, as a floating tool, allows for the weight of the anchor chain and the underwater cable to not be added to the balloon load. Moreover, the balloon cable is fastened to the buoy through a damper to attenuate nonstationary stresses on this cable arising from the motion of the buoy and the balloon, especially in the cases of a hard sea and winds.

3.3.4 Balloon Complex

As stated in Sects. 3.2 and 3.3.1, in the typical case of a large distance between the seismic epicenter at the seafloor and a seabed station, the first measurable signal of the seismic excitation in the ocean lithosphere, namely the diffusive low-frequency EM signal, propagates above the sea. A moored balloon (numbered 2 in Fig. 3.3) with a corresponding receiver, including an antenna of an appropriate size, is needed to register this signal. Indeed, this signal may be reflected by the ionosphere domain D, or it may be absorbed, attenuated, and distorted by the ionosphere spatial and temporal irregularities (say, geomagnetic storms) before reception by a satellite-borne antenna, with its rather districted sizes. So the balloon cable must be a Kevlar rope combined with a low-frequency receiving antenna, which also serves as the power supply line for the balloon-borne devices (numbered 3 in Fig. 3.3).

A scientific ultralong-duration moored balloon, filled with helium, should be located at a height of the order of one or a few kilometers, depending on the place and the intended function, but in any case before the tropopause (see Sect. 3.3.6). Problems with winds (especially hurricanes) and aircraft must be taken into account. The balloon must be fastened by a cable (additional declined guy ropes may be applied to fix the horizontal coordinates of the balloon), with a damper, to the moored buoy or to an anchor on the seafloor. In this case, additional floating tools should be applied to compensate for the weight of the additional underwater chain and cable, and to reserve the weight capacity for the instrumentation and solar panels on the balloon. Let us note that a duration of up to 100 days is possible and, a special fabric may be applied to prevent considerable leakage of helium over a longer time interval. Solar panels should be considered as additional energy sources for balloon-borne devices, taking account of the light markers needed for nighttime, according to international rules.

On the whole, similar balloons provide the possibility of much more detailed observations (i.e., at finer spatial and temporal scales) of seaquake precursory phenomena than satellites do, but over a rather more restricted area with a size of the order of a few hundred kilometers. Besides, the satellite instrumentation data are being recorded in the physical conditions of the ionosphere plasma, with its own intensive EM processes, whereas the dynamics of the EM background of the balloon-borne instrumentation are not so intense, except during thunderstorms.

So, the main applications of a moored balloon, as a part of TSUNPREDET, are:

1. Recording of seismically originated changes, including the vertical components of the field gradients along the balloon cable (see number 14 in Fig. 3.3), EM and meteorological parameters, and oceanographic data, e.g., the mean sea level and "tsunami shadows"—i.e., specific dark domains in the turbulent frontal part of tsunami waves. (Let us note that similar recordings in the medium just above the ocean, including the lower ionosphere domain D, are not directly available to satellites.)
2. Monitoring of concentrations of chemical components and aerosols of the medium near the balloon, especially the presence of new ones (e.g., radon) as a result of activation of geological structures near ocean lithosphere plate contact.
3. Monitoring of the upper atmosphere and ionosphere with a low-power ionosonde (e.g., a Lowell Digisonde, with a special transmitter and receiver) and radio-tomography signals; the latter should be directed from the balloon to the satellite (Oraevsky et al. 2000).

4. Communication with the GPS to monitor possible displacements of the moored balloon caused by winds and by day/night altitude variations, though these are expected to be minimal for an ultralong-duration moored balloon (see above).
5. Estimation of the epicenter azimuth in the case of three or more balloons with magnetometers (see Sect. 3.3.2).
6. Transmission of these multidisciplinary data in a digital form, along with data obtained by the seafloor and sea surface complexes, to a data-processing center via a satellite link or via other balloons.
7. Improvement in the validation of the satellite observational data due to points 1, 2, and 3.

For these observations and detections, different types of high-tech optics are used: spectrometers, radiometers (e.g., the Advanced Spaceborne Thermal Emission and Reflection Radiometer (ASTER) from the Jet Propulsion Laboratory, California Institute of Technology), microwave synthetic aperture radar (SAR), and vector and scalar magnetometers to provide comparisons with the corresponding satellite-borne instrumentation data and computations. Recently, there has been a variety of types of similar instrumentation with positive orbital results (e.g., Jason 1 and 2); detailed specifications are not needed here. But survey channels with ranges of 0.46–0.70 μm (visual), 6.0–7.0 μm (middle infrared (IR)), and 10.5–12.5 μm (thermal IR), with the ability to distinguish details of a size of the order of 1000 m, every 15–30 min, are needed to discover changes in oceanic surface characteristics (including color and temperature), which may be caused by a submarine seismic preparation process (Blaunstein and Hayakawa 2009; Gower 2005; Lin et al. 2010).

Use of a high-altitude balloon, generally reaching between 18 and 36 km, may be considered to obtain a more representative observation of the evolution of signals of different physical natures along the trace from a seafloor complex up to a low-altitude satellite complex and, therefore, to provide more confidence that informative prognostic signals have not been missed, as well as dismissing ones that have been generated by the investigated composed active medium without geodynamic processes.

3.3.5 *Satellite Complex*

This complex (numbered 1 in Fig. 3.3) is needed for observation and transmission of the various data recorded by the seafloor, sea surface, and balloon complexes or a whole network of them. Those data include:

- Radio-tomography
- Antenna data received by the vertical cable of the balloon
- Variations in the field gradients (see above)
- Control information on the functioning of all of the devices
- Data from the microwave radars, optic high-resolution systems, and thermal emission and reflection radiometers on the balloon and satellite

After the corresponding processing of these data—taking into account the surface roughness, images of the tsunami waves, comparison with other instrumentation data (e.g., ionosphere soundings with GPS, etc.)—the resulting prognostic estimations should be obtained (Artru et al. 2005; Calais and Minster 1998; Davies and Hartmann 1997; Gower 2005; Lin et al. 2010; Ocal et al. 1999; see also http://www.satimagingcorp.com).

Satellite electromagnetic induction methods, along with the corresponding mathematics, are effective (and provide wide coverage) in investigations of the earth's upper mantle and the electrical conductivity of its crust, but the latter is closely connected with geodynamic deformation processes, including seismogenic ones. Is it possible to extract information about seismogenic changes in conductivity under the ocean floor from recordings (of appropriate periods of signals) of the balloon- and satellite-borne vector and scalar magnetometers? Data from various sensors, including seismometers in the seafloor complexes linked to a satellite, should be used to try to clarify this prognostic problem.

Let us repeat that the data obtained by the seafloor, sea surface, balloon-borne, and satellite-borne measurement and observational instrumentation (see above) must be compared to understand the difference between (a) the distorting influences of the structure and dynamics of the ocean, atmosphere, and ionosphere; and (b) natural and seismically originated changes in different physical fields in these media (Everett et al. 2003; Hayakawa et al. 2012; Jeremy et al. 2009; Olsen 1999; Thomas et al. 2012; Usui and Omura 2007; see also http://www.esa.int/esaLP/LPswarm.html and http://www.intermagnet.org).

3.3.6 Different Configurations of the TSUNPREDET System

The minimal configuration includes a seafloor complex with batteries and a balloon complex with solar panels only.

An expanded configuration of a seafloor complex includes a set of electrodes immersed (e.g., by pivots; see Sect. 3.2) in the sedimentary layer. The electrical potentials of the electrodes can be measured relative to one of them (the reference electrode). Statistical analysis may be applied for time series of oscillations of these electrical potentials to discover changes in their statistical characteristics in the course of development of a submarine seismic preparation process. For example, coherence function values (CFVs) may be calculated for oscillations of potentials of different pairs of electrodes during consequent time intervals. It is possible that changes in these values in different frequency ranges accompany seismic preparations and may be considered signs of a seismic activation. A similar approach may be applied for oscillations of temperature and other parameters. Radio techniques for precursory signal reception have been described by Hayakawa (2015).

Configuration of the TSUNPREDET system with an additional complex, namely a sea depth complex, may be considered. This complex includes CTD (conductivity, temperature, depth) devices, (see, for example, Sherwood 2011), fastened to the anchor chain (numbered 7 in Fig. 3.3), to measure these parameters at different depths, from the seafloor boundary layer up to the ocean mixed layer. It is reasonable because the top lithosphere layer in contact with seawater, namely the sedimentary layer, is a porous medium saturated with dissolved chemicals. Some of them have become located there as a result of upward mass and heat transfer from the underlying lithosphere geological structures and are represented by their dissociated forms, i.e., charged particles (ions). The temperature field and the fields of chemical concentrations in the sedimentary layer interact with similar fields in the seawater boundary layer according to the contact conditions of continuity of the field values and their flows (e.g., the heat flow) by moving over the contact surface. As a result, changes in the fields in the sedimentary layer are accompanied by changes in the fields in the seawater boundary layer. In fact, a seismogenic precursory process deforms the porous structure and changes the temperature field and the fields of dissociated chemical concentrations, i.e., charged particles (the PDEs of a proper mathematical model are omitted), in the sedimentary layer. These disturbances in the temperature field and the fields of charged particle concentrations will be transferred, in the course of the precursory process, from the top of the sedimentary layer into the seawater depth, according to the field contact conditions. These heat and mass disturbances will propagate there in different directions, including upward motion. Changes of temperature and electrical conductivity (because of arising charged particles) will be registered by the CTD devices, beginning from the lower ones and then, after a delay, by ones located higher. The time delay depends on the vertical velocities of the temperature disturbance propagation and the charged particle diffusion.

In the case of a TSUNPREDET system located near a submarine fault or lithosphere plate contact, an earlier reaction of the lower CTD devices, as compared with the reaction of the higher ones, will be expressed in a more distinctive way due to intensive heat and mass transfer (convection) from the depth of similar structures by their seismic activation. In any case, an anticipating reaction of the lower CTD devices may be interpreted as a sign of a seismic preparation process.

A configuration may be applied in which a balloon complex is linked with a net of seafloor complexes and transmits information from them and itself to a processing center without a satellite, taking account of the fact that the maximum possible distance for a shortwave balloon-borne transmitter located lower than the tropopause (this case is described in Sect. 3.3.3) is of the order of a few hundred kilometers, and a properly directed TSUNPREDET chain may be used, depending on the configuration of the shore being protected. In the general case of a detection system located far from the shore (e.g., depending on the contact of the ocean lithosphere plates), a satellite link should be applied. A full regional tsunami precursor detection array should be represented by a suitable combination of local and general detection systems, taking account of the fact that a submarine seismic source may be located near the shore or far from it.

3.3.7 Conclusion

On the basis of the main physical principles and geophysical data, we have formulated a mathematical model of the seismo-hydro-electromagnetic-temperature geophysical field interaction (in the presence of gravity) arising in the lithosphere–ocean–atmosphere medium as the result of a weak seismic excitation (see Sect. 3.2) in the upper mantle under the ocean. The magnitudes of the computed seismic, electromagnetic, temperature, and hydrodynamic wave amplitudes, and the main frequencies and velocities, including the delay in regard to the beginning of a submarine earthquake, are of the orders of magnitude observed and published by other authors. According to the theory developed and the measurements, a scheme for a multidisciplinary vertically distributed (from the seafloor up to the ionosphere) tsunami precursors detection system is demonstrated.

Appendix A: SHEMTI Equations

String–Diffusion Structure of the SHEMTI Model

Weak seismic excitations (see Sects. 3.1 and 3.2), supplying energy for the SHEMTI process of tsunami precursor formation, may be described by operator $\mathcal{H}$ of the linear dynamical theory of elasticity

$$\mathcal{H}\mathbf{u} \equiv \rho \frac{\partial^2 \mathbf{u}}{\partial t^2} - \mathrm{Div}\widehat{\sigma},$$

$$\widehat{\sigma} = (\sigma_{ij})_{i,j=1}^{i,j=3}, \quad \sigma_{ij} = 2\mu(\mathbf{x})\varepsilon_{ij} + \lambda(\mathbf{x})\delta_{ij} div\mathbf{u},$$

$$\varepsilon_{ij} = \frac{1}{2}\left(\frac{\partial u_i}{\partial x_j} + \frac{\partial u_j}{\partial x_i}\right), i,j = 1,2,3$$

So, the stress tensor σ_{ij} is supposed to obey the Hook strain–stress relation with the Cauchy strain tensor ε_{ij} and the Lame elastic parameters μ and λ, and Div is used for the divergence of the stress tensor:

$$\mathrm{Div}\widehat{\sigma} = \left((\mathrm{Div}\widehat{\sigma})_1, (\mathrm{Div}\widehat{\sigma})_2, (\mathrm{Div}\widehat{\sigma})_3\right)^{\mathrm{tr}}, (\mathrm{Div}\widehat{\sigma})_i = \sum_{j=1}^{3} \frac{\partial \sigma_{ij}}{\partial x_j}, \quad i = 1,2,3$$

In these notations, the system of partial differential equations (PDEs) of the dynamic theory of elasticity takes the form:

$$\mathcal{H}\mathbf{u} = \rho\mathbf{g}, \tag{3.1}$$

where:

$$\mathbf{u} = (u_1(t,\mathbf{x}), u_2(t,\mathbf{x}), u_3(t,\mathbf{x}))^{\mathrm{tr}}, \mathbf{x} = (x_1, x_2, x_3) = (x, y, z), \mathbf{g} = (g_1, 0, 0)^{\mathrm{tr}}$$

are the vectors of the elastic displacements, coordinates, and acceleration of gravity, respectively; indexes 2 and 3 are for the horizontal directions and 1 is for the vertical one; and $\rho = \rho(\mathbf{x})$ denotes the density.

Operator $\mathcal{H}$ is determined formally, i.e., without functional spaces (Tahery 2015), because this chapter is not a mathematical one. This operator belongs to the known hyperbolic type (the operator of a *string* is the simplest example): the energy of deformation processes described by similar operators does not disappear. But this is not true for rocks, because temperature (T) and EM fields are generated by deformations (see Sects. 3.1 and 3.2). Therefore, operator $\mathcal{H}$ should be combined with operator $\mathcal{P}$ of a parabolic type (see Ladyzhenskaya et al. 1967), describing dissipative processes.

In the case of seismic EM generation in conductive rocks with a minimal electrical conductivity of the order of 0.0001 S/m, these processes include propagation of a low-frequency (from 0.01 up to 100 Hz) diffusive field of magnetic induction:

$$\mathbf{B}(t,\mathbf{x}) = (B_1(t,\mathbf{x}), B_2(t,\mathbf{x}), B_3(t,\mathbf{x}))^{\mathrm{tr}}$$

and a thermal field with deviation $\theta(t, \mathbf{x})$ of the temperature $\Theta(t, x)$ from the initial temperature θ_0 existing at $t = 0$. Let us denote these dissipative processes by $\mathbf{w}(t, \mathbf{x})$:

$$\begin{aligned}\mathbf{w}(t,\mathbf{x}) &\equiv (\mathbf{H}(t,\mathbf{x}), \theta(t,\mathbf{x}))^{\mathrm{tr}}, \mathbf{H}(t,\mathbf{x}) = (H_1(t,\mathbf{x}), H_2(t,\mathbf{x}), H_2(t,\mathbf{x}))^{\mathrm{tr}} \\ &= (\mu_0)^{-1}\,\mathbf{B}(t,\mathbf{x}),\end{aligned}$$

where μ_0 is the magnetic permeability of vacuum. The medium is supposed to be uniform in regard to magnetic permeability, and its value is μ_0.

So, the diffusion in the medium under consideration is described by the parabolic vector equation (i.e., the system of PDEs of the parabolic type):

$$\mathcal{P}\mathbf{w} = \mathbf{q}, \tag{3.2}$$

where:

$\mathbf{q} = (0, 0, q(\mathbf{x}))^{\mathrm{tr}}$ is the density of the power of heat sources (e.g., the radioactivity of the rocks),

$$\mathcal{P}\mathbf{w} \equiv \begin{pmatrix} \dfrac{\partial \mathbf{B}}{\partial t} + \mathrm{rot}\left(\sigma_e^{-1}(\mathbf{x})\mathrm{rot}\mathbf{H}\right) \\ \rho(\mathbf{x})c_V\dfrac{\partial \vartheta}{\partial t} - \mathrm{div}(\kappa(\mathbf{x})\mathrm{grad}\vartheta) \end{pmatrix}$$

Here, the first line, which is a vector with three components, determines the operator of magnetic diffusion (the low-frequency displacement currents are negligible as compared with the ones of conductivity) by electrical conductivity σ_e, and the second line determines the scalar diffusion operator of heat transfer by heat conductivity κ; heat capacity is denoted by c_V.

Contrary to observations in seismically active regions, the classical hyperbolic and parabolic vector Eqs. (3.1) and (3.2) are not able to describe generation of magnetic (**H**) and temperature (θ) field disturbances by seismic excitation of a medium, i.e., magnetic and temperature precursors of earthquakes (see Sect. 3.1).

Along with the above assumptions, let us suppose that polarization (magnetic, electrical) and thermoelectric processes are negligible and a change in the physical characteristics of rocks is impossible under the influence of a weak precursory excitation.

For the conditions mentioned, we apply the magneto-thermo-elasticity (MTE) theory (Maugin 1988) to describe seismo-EM-thermal interaction in the lithosphere. Eq. (3.1) is replaced by (3.3) and Eq. (3.2) by (3.4). So, a coupled system is obtained:

$$\mathcal{H}\mathbf{u} + \mathcal{A}(\mathbf{w}) = \rho\mathbf{g} \tag{3.3}$$

$$\mathcal{P}\mathbf{w} + \mathcal{B}(\mathbf{u}, \mathbf{w}) = \mathbf{q} \tag{3.4}$$

with the interaction operators in the form of an additional force $\mathcal{A}(\mathbf{w})$ and field source $\mathcal{B}(\mathbf{u}, \mathbf{w})$,

$$\mathcal{A}(\mathbf{w}) \equiv \mathrm{grad}(\beta(\mathbf{x})\vartheta) + \mathbf{B} \times \mathrm{rot}\mathbf{H}$$

$$\mathcal{B}(\mathbf{u}, \mathbf{w}) \equiv \begin{pmatrix} -\mathrm{rot}[\mathbf{v} \times \mathbf{B}] \\ \beta(\mathbf{x})\vartheta_0\mathrm{div}\mathbf{v} - \sigma_e^{-1}(\mathbf{x})(\mathrm{rot}\mathbf{H})^2 \end{pmatrix}$$

where:

$\mathbf{v} = \partial\mathbf{u}/\partial t$, $\mathbf{v}^2 = v_1^2 + v_2^2 + v_3^2 << c^2$, c is the light velocity in a vacuum and the inequality represents the known assumption of slow motion, $\beta = (2\mu + 3\lambda)\alpha_\theta$, α_θ is the coefficient of thermal expansion, $\alpha_\theta = 5 \times 10^{-6}\ \mathrm{K}^{-1}$. The first term in the expression for interaction operator $\mathcal{A}(\mathbf{w})$ describes the thermo-stress and the second one originates from the Lorentz force. The first (vector) line in the expression for interaction operator $\mathcal{B}(\mathbf{u}, \mathbf{w})$ arises by applying the operator rot to the Maxwell equation (without displacement currents; see above): rot $\mathbf{H} = \mathbf{j}$, $\mathbf{j} = \sigma_e(\ \mathbf{E}+[\mathbf{v} \times \mathbf{B}]\ )$, $\mathbf{j}$ is the density of the conductivity current in a slowly moving medium, and $\mathbf{E}$ is the electrical intensity. The first and second terms of the second (scalar) line of the operator $\mathcal{B}(\mathbf{u}, \mathbf{w})$ describe the heat production due to motion and electrical currents, respectively.

System (3.3), (3.4) of PDEs of the seismo-EM-thermal interaction is not of a classical type, because of the different, namely string and diffusion, physical natures of processes **u**, **w** and, correspondingly, different mathematical types of operators $\mathcal{H}$, $\mathcal{P}$ (see above). Let us note:

$$\mathcal{A}(\mathbf{w}) \to \mathbf{0}, \mathcal{B}(\mathbf{u}, \mathbf{w}) \to \mathbf{0} \quad \text{by} \quad \mathbf{w} \to \mathbf{0}, \mathbf{u} \to 0$$

In other words, if the physical fields under consideration are weak enough, then the string–diffusion (S.-D.) system of PDEs (partial differential equations) takes the form of Eqs. (3.1), (3.2) of the classical types. By going over to the seismo-hydro-EM-thermal interaction in the lithosphere–ocean medium, we apply the oceanology model of "shallow water" with its hyperbolic operator of motion (e.g., see Landau and Lifshits 1986), instead of one from the theory of elasticity described above. So, the resulting mathematical structure of the three-dimensional (3D) model of the SHEMTI process is of the S.-D. type, i.e., represented by hyperbolic and parabolic operators coupled with field interaction operators (see above) and contact conditions at the lithosphere–ocean interface. To reduce the number of written equations, we describe the full S.-D. form of the SHEMTI model for the two-dimensional (2D) case only (see next section) because the 3D case was computed for the seismo-EM interaction only.

Initial Boundary Value Problem of the 2D SHEMTI Model

We denote the seismic disturbance of the "scalar magnetic potential" as A, that of the magnetic field intensity as $\mathbf{H} = (H_1, H_2)^{tr}$, and that of magnetic induction as $\mathbf{B} = (B_1, B_2)^{tr}$, $H_1 = -\partial A(t, x_1, x_2)/\partial x_2$, $H_2 = \partial A(t, x_1, x_2)/\partial x_1$.

Also, we use the following 2D thermo-mechanical notations: $\lambda = \lambda(x_1, x_2)$ and $\mu = \mu(x_1, x_2)$ are parameters of Lame, $\mathbf{u} = (u_1(t, x_1, x_2), u_2(t, x_1, x_2))^{tr}$ is the elastic displacement at point (x_1, x_2) at time t, $v_i = \partial u_i/\partial t$, $i = 1, 2$, are the components of velocity; $\boldsymbol{g} = (g,0)^{tr}$ is the vector of acceleration of gravity. The evolution of the interacting thermo-mechanical and EM fields in the lithosphere part of the medium according to the 2D version of the MTE model (3.3), (3.4) obeys:

$$\begin{gathered}
\rho \frac{\partial^2 u_i}{\partial t^2} = \frac{\partial}{\partial x_k}\left(\mu\left(\frac{\partial u_i}{\partial x_k} + \frac{\partial u_k}{\partial x_i}\right)\right) + \frac{\partial}{\partial x_i}\left(\lambda \frac{\partial u_k}{\partial x_k}\right) - \frac{\partial(\beta\vartheta)}{\partial x_i} + \rho g_i, \quad i = 1,2 \\
\frac{\partial A}{\partial t} = \frac{1}{\mu_0 \sigma_e}\Delta A - v_k \frac{\partial A}{\partial x_k} + v_2 H_{10} - v_1 H_{20} \\
\rho c_V \frac{\partial \vartheta}{\partial t} = \operatorname{div}(\kappa \cdot \operatorname{grad}\vartheta) - \beta\vartheta_0 \operatorname{div}\mathbf{v} + \frac{1}{\sigma_e}(\Delta A)^2 + q
\end{gathered} \tag{3.5}$$

where $\rho = \rho(x_1, x_2)$ is the density, $\sigma_e = \sigma_e\ (x_1, x_2)$ is the electrical conductivity, $\kappa = \kappa(x_1, x_2)$ is the thermal conductivity, $q = q(x_1, x_2)$ is the density of the power of heat sources, and $\mathbf{H}_0 = (H_{10,}\ H_{20})^{tr}$ is the stationary (unperturbed) magnetic field; here and below there is a summation over repeated subscripts which take at 1 and 2, subscript 1 corresponds to the vertical direction and 2 to the horizontal one. The EM term is omitted in the equations of motion, $i = 1, 2$, because of its negligible influence, according to our calculations.

At the sea bottom the normal component v_n of the elastic velocity $\mathbf{v}$ equals the normal component w_n of the water velocity $\mathbf{w} = (w_1, w_2)^{tr}$:

$$(\mathbf{v} \cdot \mathbf{n}) = (\mathbf{w} \cdot \mathbf{n}) \tag{3.6}$$

where $\mathbf{n}$ and $\boldsymbol{\tau}$ (see Eq. (3.7)) are the local normal and tangent to the sea bottom (the normal is directed inwards the water). Then, the normal component of force f_n acting from the sediment layer equals the water pressure p at this point, while the tangent component of this force f_τ vanishes because we neglect shear stresses in the water:

$$f_n = \sigma_{ij} n_i n_j = p, \quad f_\tau = \sigma_{ij} \tau_i n_j = 0, \tag{3.7}$$

where the Hook's relation is used in the form:

$$\sigma_{ij} \equiv \mu\left(\frac{\partial u_i}{\partial x_j} + \frac{\partial u_j}{\partial x_i}\right) + \delta_{ij}\lambda \frac{\partial u_k}{\partial x_k} \tag{3.8}$$

We describe the water dynamics in the framework of the nonlinear "shallow water" theory (e.g., Landau and Lifshits (1986)), which operates the horizontal velocity w_2 and elevation of the sea surface $\eta(t, x_2)$:

$$\begin{gathered} \frac{\partial}{\partial t}(\eta + h_u) + \frac{\partial}{\partial x_2}((\eta + h)w_2) = 0, \\ \frac{\partial w_2}{\partial t} + w_2 \frac{\partial w_2}{\partial x_2} + g\frac{\partial \eta}{\partial x_2} = 0 \end{gathered} \tag{3.9}$$

The sea bottom is a moving line:

$$x_1 = h(t, x_2) = h_0(x_2) + h_u(t, x_2),$$

where $h_0(x_2)$ is the stationary profile and $h_u(t, x_2)$ is its perturbation due to elastic deformations caused by seismic excitation (3.1) in the upper mantle. In seawater, we neglect the seismic disturbance of the temperature field and the Lorentz force, so the interaction operator $\mathcal{A} = 0$ there.

Now let us consider the seismo-hydro-EM interaction in the ocean–atmosphere system. The potential A in the water obeys:

$$\frac{\partial A}{\partial t} = \frac{1}{\mu_0 \sigma_e} \Delta A - w_k \frac{\partial A}{\partial x_k} + w_2 H_{10} - w_1 H_{20} \tag{3.10}$$

The vertical component w_1 of the hydrodynamic velocity is calculated from w_2 and elastic velocity using Eqs. (3.7), (3.9), and the incompressibility condition:

$$\frac{\partial w_1}{\partial x_1} + \frac{\partial w_2}{\partial x_2} = 0 \tag{3.11}$$

Electrical conductivity is assumed to be continuous across the sea bottom (i.e., a thin transition layer is used in the computations).

The equation for A above the sea (where $x_1 < 0$) and the conjugation conditions at the sea surface $x_1 = 0$ are:

$$\begin{aligned} &\Delta A = 0 \quad \text{for } x_1 < 0, \quad 0 < x_2 < L_2, \quad t > 0; \\ &A|_{x_1=+0} = A|_{x_1=-0}, \quad \left.\frac{\partial A}{\partial x_1}\right|_{x_1=+0} = \left.\frac{\partial A}{\partial x_1}\right|_{x_1=-0} \end{aligned} \tag{3.12}$$

The *initial conditions* were chosen so that before the moment $t = 0$, the system was at rest; thus, the water velocities and elevation of the sea surface vanish, the elastic displacements equal the stationary configuration under the gravity force, and the initial magnetic field is uniform in space: $H_{10} = 30$, $H_{20} = 35$ A/m. For all fields, the normal derivative vanishes at the left ($x_2 = 0$) and right ($x_2 = L_2$) boundaries. The remaining "outer" boundary conditions are (see L_D, L_I, and other input in Appendix B):

$$\begin{aligned} &\partial A/\partial x_1|_{x_1=-L_D} = \partial A/\partial x_1|_{x_1=L_1} = 0, \\ &\vartheta|_{x_1=h_0(x_2)} = 0, \quad \partial\vartheta/\partial x_1|_{x_1=L_1} = 0, \\ &\mathbf{u}|_{x_1=L_1} = \mathbf{u}_0(t, x_2) \end{aligned} \tag{3.13}$$

This approach to the theory of the seismo-hydro-EM-temperature processes, including the existence and uniqueness theorems for the solution of the initial boundary value problem of the system of PDEs with the coupled operators of the hyperbolic and parabolic types, e.g., (3.5–3.13), and the corresponding discrete approximations, is a further development of Novik (1995), Novik et al. (2001), Novik and Ershov (2001), Ershov et al. (2001, 2006), and Ershov and Novik (2009). Other approaches to the seismo-electromagnetics and bibliography may be found in the references.

Appendix B: Data Applied to the Calculations in Sect. 3.2

We consider a 2D problem of seismic generation and propagation of elastic, hydrodynamic, EM and temperature (generated by dissipation in rocks) field disturbances in a domain including a vertical cross-section of the central part of the Sea of Japan down to the upper mantle at a depth of $L_1 = 37$ km below the sea level, the sea, and the atmosphere up to the lower boundary of the ionosphere domain D at the height $x_1 = -L_D$, $L_D = 70$ km.

The cross-section, with a length of 425 km, runs from the Bay of Peter the Great through the Pervenets Rise and the Yamato Basin, and ends at the North Yamato Rise. The coordinates $x = (x_1, x_2)$ are chosen so that x_1 is directed downward with $x_1 = 0$ at the sea surface, and x_2 is directed horizontally, $0 \le x_2 \le L_2 = 425$ km.

The seawater and the sedimentary, granite, basaltic, and upper mantle layers are denoted by *W*, *S*, *G*, *B*, and *M*, respectively (Fig. 3.4 and Table 3.1). The heat capacity is $c_V = 660$ J/(kg × K) everywhere and the magnetic permeability is $\mu_e = \mu_0 = 4\pi \times 10^{-7}$ (in SI units). The components H_{10} and H_{20} of the stationary geomagnetic field H_0, $\mathbf{H}_0 = (H_{10}, H_{20})^{tr}$, are supposed to be uniform in the considered domain.

To model the seismic excitation due to the arrival of a seismic wave from a deeper source at the lower external boundary $x_1 = L_1 = 37$ km, we use the vector $\mathbf{u}_0 = (u_{01}, u_{02})^{tr}$ as a nonstationary boundary condition for elastic displacements $u_1(t, x_1, x_2)$ and $u_2(t, x_1, x_2)$:

$$u_1(t, L_1, x_2) = u_{01}, u_2(t, L_1, x_2) = u_{02}$$

where:

$$u_{01} = a_1 \times te^{-bt-\alpha(x_2-x_{20})^2} \cos(\omega t) \cos \frac{2\pi x_2}{d},$$
$$u_{02} = a_2 \times te^{-bt-\alpha(x_2-x_{20})^2} \sin(\omega t) \sin \frac{2\pi x_2}{d} \tag{3.14}$$

Here: $a_1 = 7$ cm, $a_2 = 3.5$ cm, $b = 1/3$ s^{-1}, $\alpha = 5 \times 10^{-11}$ m^{-2}, $0 < t < T$, $\omega = 5/3$ s^{-1}, $x_{20} = L_2/2 = 212.5$ km, $d = 150$ km, $T = 20$ s.

Table 3.1 Physical parameters of the medium: values (in SI units) of density ρ, elastic wave velocities v_p and v_s, electrical and thermal conductivities σ_e and κ, and densities of heat flow q

	ρ (kg/m^3)	v_p (m/s)	v_s (m/s)	σ_e (S/m)	κ (W/m × K)	q (μW/m^3)
W	1000	–	–	3.5	–	–
S	2500	5000	3000	0.005	1.3	1.2
G	2700	5800	3200	0.001	2.5	0.9
B	2900	6300	4000	0.005	2.1	0.4
M	3300	7200	4800	0.020	3.2	0.4

B basaltic layer, *G* granite layer, *M* upper mantle layer, *S* sedimentary layer, *W* seawater

The duration of this model seismic excitation (varied in different simulations) is several seconds, the amplitude is about 5 cm, the main frequency is about several tenths of a hertz (the spectrum of u_{02} is the same), and the excitation decays with time and far from the center of the domain.

Let us formulate a few remarks about the dependence of the characteristics of the computed SHEMTI process on the variations of the data in its string–diffusion model.

Our simulation algorithms allow us to compute evolution of fields under a wide class of initial and boundary conditions, including volumetric a seismic excitation instead of a boundary excitation. Indeed, a computed field configuration for some moment may be considered as the initial one for later moments. In the frame of the 2D string–diffusion model (3.5–3.13) of the SHEMTI process, we have varied the parameters of the excitation and other conditions at the outer boundaries $x_1 = L_1$, $x_1 = -L_D$ and $x_2 = 0$, $x_2 = L_2$, as well as the characteristics of the medium (taking into account the geophysical and petrophysical restrictions), but the main features of the evolution of the model seismic perturbations of fields, especially the relation between seismic excitations in the earth depth under the seafloor and EM signals at the sea surface and above, have remained the same. These computations show that the values of the parameters of the seismic excitation (i.e., the nonstationary condition at the lower boundary of the domain considered) are of the most importance, as compared with other data, in forming (a) the seismo-EM process in the lithosphere part of the medium (i.e., the energy supply of the tsunami), and (b) EM signals above the sea. Our 3D computations of the seismo-EM process with the string–diffusion model (3.1–3.4) for the case of a geological structure with a weak dependence of its properties on one of the horizontal coordinates (lineament, e.g., a domain near a central cross-section of a structure) support the 2D computations; for example, the computed long EM wave inherits the time and space characteristics of the seismic wave for both 2D and 3D geometries. Also, the hydrodynamic component of the SHEMTI process has been computed with the help of the hydro-acoustic operator instead of the "shallow water" one (both are of the hyperbolic type). The pictures of the computed SHEMTI processes are similar in terms of the main details.

References

Artru, J., Ducic, V., Kanamory, H., et al. (2005). Ionosphere detection of gravity waves induced by tsunamis. *Geophysical Journal International, 160*, 840–848.

Basseville, M., & Benveniste, A. (Eds.). (1986). *Detection of abrupt changes in signals and dynamical systems*, *Lecture notes in control and information sciences* (Vol. 77). Berlin: Springer.

Belov, S. B., Migunov, N. I., & Sobolev, G. A. (1974). Magnetic effect of the strong earthquakes at Kamchatka. *Magnetism and Aeronomy, 14*, 380–382 (in Russian).

Beransoli, L., Favali, P., & Smriglio, G. (Eds.). (2002). *Science–technology synergy for research in the marine environment, challenges for the XXI century*, *Developments in marine technology* (Vol. 12). Amsterdam: Elsevier.

Blaunstein, N., & Hayakawa, M. (2009). Short-term ionospheric precursors of earthquakes using vertical and oblique ionosondes. *Physics and Chemistry of the Earth, Parts A/B/C, 34*(6–7), 496–507. Special issue, Electromagnetic phenomena associated with earthquakes and volcanoes, Edited by M. Hayakawa, J. Y. Liu, K. Hattori, and L. Telesca.

Calais, E., & Minster, J. B. (1998). GPS, earthquakes, the ionosphere, and the space shuttle. *Physics of the Earth and Planetary Interiors, 105*, 167–181.

Casey, K., Lim, A., & Dozier, G. (2008). A sensor network architecture for tsunami detection and response. *International Journal of Distributed Sensor Networks, 4*(1), 28–43.

Constable, S. C., Orangez, A. S., Hoversten, G. M., & Morrison, H. F. (1998). Marine magnetotellurics for petroleum exploration. Part I: A sea-floor equipment system. *Geophysics, 63*, 816–825.

Davies, K., & Hartmann, G. K. (1997). Studying the ionosphere with the global positioning system. *Radio Science, 32*, 1695–1698.

Ershov, S., & Novik, O. (2009). *Elements of the tsunami precursors' physics: Seismo-hydro-electromagnetics*. In Proceedings (CD) of the 6th international marine electromagnetics conference (MARELEC 2009) (pp. 122–131), Stockholm.

Ershov, S., Mikhaylovskaya, I., & Novik, O. (2001). Electromagnetic signals of seismic excitation of geological structures beneath a sea bottom. *Physics and Chemistry of the Earth, 26*, 761–768.

Ershov, S., Mikhaylovskaya, I., & Novik, O. (2006). Theory of EM monitoring of sea bottom geothermal areas. *Journal of Applied Geophysics, 58*, 330–350.

Everett, M. E., Constable, S., & Constable, C. G. (2003). Effects of near-surface conduction on global satellite induction responses. *Geophysical Journal International, 153*, 277–285.

Gohberg, M. B., Krylov, S. M., & Levshenko, V. T. (1991). Electromagnetic field of the earthquake focus. *Transactions (Doklady) of the Russian Academy of Sciences, 308*, 1–3. (Translated from Russian, Doklady Acad. Nauk USSR, 308, 62–65, 1989).

González, F. I., Bernard, E. N., Meinig, C., Eble, M., Mofjeld, H. O., & Stalin, S. (2005). The NTHMP tsunameter network. *Natural Hazards, 35*, 25–39.

Gower, J. (2005). Jason 1 detects the 26 December 2004 tsunami. *EOS, Transactions, American Geophysical Union, 86*(4), 37–38.

Hayakawa, M. (Ed.). (2003). *Atmospheric and ionospheric electromagnetic phenomena associated with earthquakes*. Tokyo: Terrapub.

Hayakawa, M. (2015). *Earthquake prediction with radio techniques*. Hoboken: Wiley.

Hayakawa, M., Hobara, Y., Yasuda, Y., Yamaguchi, H., Ohta, K., Izutsu, J., et al. (2012). Possible precursor to the March 11, 2011, Japan earthquake: Lonospheric perturbations as seen by subionospheric very low frequency/low frequency propagation. *Annals of Geophysics, 55*(1), 95–99. https://doi.org/10.4401/ag-5357 (Special issue on earthquake precursors).

Hayman, N. W., Bach, W., Blackman, D., Christeson, G. L., Edwards, K., Haymon, R., et al. (2010). Future scientific drilling of oceanic crust. *EOS, Transactions, American Geophysical Union, 91*(15), 133–134.

Iyemori, T., Kamei, T., Tanaka, Y., Takeda, M., Hashimoto, T., Araki, T., et al. (1996). Co-seismic geomagnetic variations observed at the 1995 Hyogoken–Nanbu earthquake. *Journal of Geomagnetism and Geoelectricity, 48*, 1059–1070.

Jeremy, N., Thomas, J. N., Love, J. J., Johnston, M., & Yumoto, K. (2009). On the reported magnetic precursor of the 1993 Guam earthquake. *Geophysical Research, 36*, L16301. https://doi.org/10.1029/2009GL039020.

Ladyzhenskaya, O. A., Solonnikov, V. A., & Uraltseva, N. N. (1967). *Linear and quasilinear equations of parabolic type*. Moscow: Izdat. ("Nauka". (1968). Translations of Mathematical Monographs (English trans; Vol. 23). Providence: American Mathematical Society, 1967).

Landau, L. D., & Lifshits, E. M. (1986). Course of theoretical physics. In *Fluid mechanics* (Vol. VI). Cambridge: Academic Press (Translated from Russian).

Lin, F. C., Nakornphanom, K., Sookhanaphibarn, K., & Lursinsap, C. (2010). A new paradigm for detecting tsunamis by remote sensing. *International Journal of Geoinformatics, 6*(1), 19–30.

Lin, J. C., Brunner, D., & Gerbig, C. (2011). Studying atmospheric transport trough Lagrangian models. *EOS, Transactions, American Geophysical Union, 92*(21), 177–178.

Maugin, G. A. (1988). *Continuum mechanics of electromagnetic solids*. Amsterdam: North-Holland.

Novik, O. B. (1995). The Galerkin method for a three-dimensional nonlinear system of magnetothermo-elasticity equations. *Transactions (Doklady) of the Russian Academy of Science, 335A*(3), 15–21. (Translated from Russian, Doklady Rossiyskoy Akademii Nauk, 1994, 334, 100–102).

Novik, O. B., & Ershov, S. V. (2001). *Electromagnetic and temperature signals from the earth's depth*. Scientific monographs of federal program of integration of high education and fundamental science, Moscow (in Russian).

Novik, O. B., Ershov, S. V., & Mikhaylovskaya, I. B. (2001). *Electromagnetic signals in atmosphere caused by dynamic processes in ocean lithosphere*. In Proceedings of the 15th European space agency symposium on rocket and balloon programs and related research (pp. 369–373), Biarritz, 28–31 May 2001.

Ocal, E., Piatanesi, A., & Heinrich, P. P. (1999). Tsunami detection by satellite altimetry. *Journal of Geophysical Research, 104*, 599–615.

Olsen, N. (1999). Induction studies with satellite data. *Surveys in Geophysics, 20*, 309–340.

Oraevsky, V. N., Ruzhin, Yu. Ya., & Shagemuratov, I. I. (2000). *Anomalies of the ionosphere TEC above Turkey*. In International Wraclaw symposium on electromagnetic compatibility (EMC 2000), proceedings (pp. 509–511).

Ruzhin, Y., & Nomicos, C. (2007). Radio VHF precursors of earthquakes. *Natural Hazards, 40*, 573–583.

Sherwood, C. (2011, November/December). Mechanical arm + internet = realtime profilers of particles near the seafloor. *Sound Waves Monthly Newsletter, USGS*. http://soundwaves.usgs.gov/2011/11/fieldwork2.html. Accessed 08 January 2019.

Siripunvaraporna, W., Egbert, G., Lenburyc, Y., & Uyeshimad, M. (2005). Three-dimensional magnetotelluric inversion, data-space method. *Physics of the Earth and Planetary Interiors, 150*, 3–14.

Tahery, A. (2015). *Function spaces and partial differential equations*. Oxford: Oxford University Press.

Thomas, J. N., Love, J. J., Komjathy, A., Verkhoglyadova, O. A., Butala, M., & Rivera, N. (2012). On the reported ionospheric precursor of the 1999 Hector Mine, California earthquake. *Geophysical Research Letters, 39*, L06302.

Toh, H., Goto, T., & Hamano, Y. (1998). A new seafloor electromagnetic station with an Overhauser magnetometer, a magnetotelluric variograph and an acoustic telemetry modem. *Earth, Planets and Space, 50*, 895–903.

Usui, H., & Omura, Y. (Eds.). (2007). *Advanced methods for space simulations*. Tokyo: Terrapub.

Chapter 4
Space Weather

Abstract In this chapter, we briefly describe the structure and dynamics of the plasma of the solar wind, which supplies solar energy to the plasmas of the earth's geospheres, namely the magnetosphere and ionosphere. Then, geomagnetic storms, which develop there and disturb the physical fields of the underlying earth atmosphere, are discussed. We use the qualitative approach of the US Space Weather Prediction Center (SWPC), the National Oceanic and Atmospheric Administration (NOAA), Spacepedia, Wikipedia, and the National Aeronautic and Space Administration (NASA) to discuss these subjects, i.e., systems of partial differential equations (PDEs; e.g., ones describing magneto-hydrodynamics) are not applied in this chapter, though the processes described by them include most of the related physics. (We note that systems of PDEs are applied and/or investigated by us numerically and mathematically in Chaps. 1, 2, 3, 9, 10, and 11 of this book.)

The influences of geomagnetic storms (estimated by the planetary geomagnetic index K_p, which is described in this chapter) on the human brain and heart are considered in Chaps. 5, 6, 7, and 8 of this book.

Keywords Plasma · Solar wind · Magnetosphere · Ionosphere · Geomagnetic storm

4.1 The Solar Wind

Ludwig Biermann, trying to understand why the tail of a comet is directed outward from the sun, discovered (in 1951) that the sun emits a giant flow of plasma containing mostly electrons, protons, and alpha particles within the embedded interplanetary magnetic field (here we do not discuss the question as to who first wrote about such a concept). This flow is now known as the solar wind. It is emitted from the corona (the solar upper atmosphere) and interacts with all parts of the solar system, including the earth. The impact of the solar wind on the earth's biosphere could be pernicious because the energy of the solar wind's particles may be of the order of 10 keV.

O. Novik et al., *Electromagnetic Geophysical Fields*,
https://doi.org/10.1007/978-3-319-98461-2_4

Fig. 4.1 Spatial structure of the solar wind

But the earth's magnetic field, namely the part of it that is located in the earth's magnetosphere (see Sect. 4.2) on its dayside, deflects most of the charged solar particles by means of the Lorentz force. Accordingly, the domain in space between the sun and the earth that is occupied by the earth's magnetic field is restricted by the solar wind pressure on the magnetosphere. The solar wind plasma is nonuniform: the solar wind from the coronal holes near the poles has a speed of 500–800 km/s, whereas the wind from the equatorial coronal holes has a speed of 400 km/s.

The planets orbit around the sun on the equatorial plane. The period of the rotation of the sun itself is 27 days. This rotation makes the structure of the solar wind plasma rather composed in regard to the spatial and temporal distribution of its density and speed (Fig. 4.1). In any case, the fast portion of the high-density solar wind is the initial reason for the strong geomagnetic storms that occur on earth. (Delamere et al. 2015; Meyer-Vernet 2007; Miralles and Almeda 2011; Waltz 2015).

4.2 The Earth's Magnetic Field and the Magnetosphere

The magnetosphere of the earth (Fig. 4.2) is the domain of space around the earth where its magnetic field is dominant, as compared with the interplanetary field, i.e., the magnetic field of interplanetary space.

The magnetosphere is formed as a result of interaction of the solar wind with the magnetic field of the earth. The latter is generated in the molten outer core region of the earth's interior. The convective currents in the molten iron there generate electrical currents and these, according to the first partial differential equation (PDE) of Maxwell (Ampère's law), generate the magnetic field; the process is supported by geothermal energy (the interesting PDE problems of the theory of this natural process, known as the *geodynamo*, will not be considered here). At the surface of the earth the magnetic field induction ranges from 25 to 65 μT. Far away from the described place of generation, the earth's magnetosphere, and therefore its magnetic field domain, are restricted by the solar wind pressure (see Sect. 4.1).

The hypothesis that the earth is a giant magnet was formulated by William Gilbert, the Queen of England's doctor, more than four centuries ago. It is now

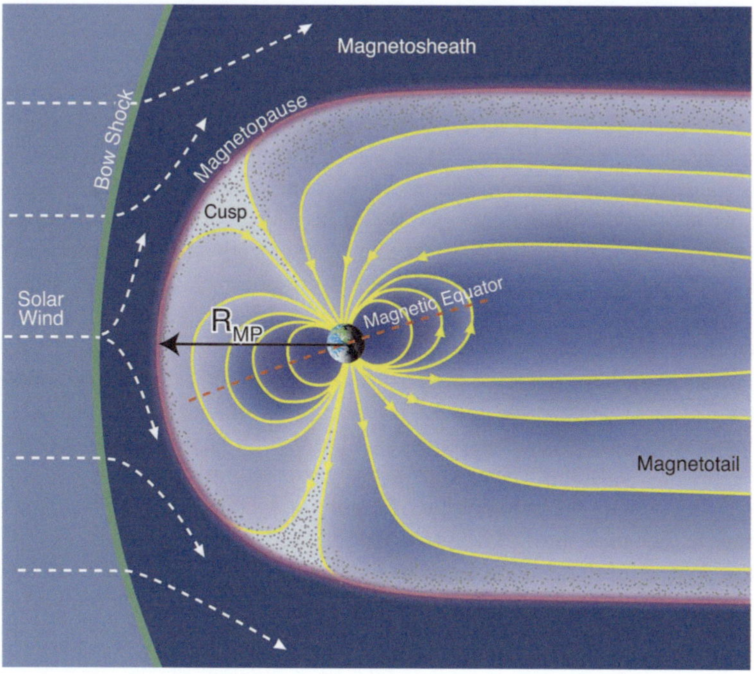

Fig. 4.2 The magnetosphere of the earth

known, after investigations in physics and mathematics (e.g., by Gauss) that the earth's magnetic field geometry is quite composed but rather roughly may be approximated by the geometry of the field of a magnetic dipole.

Because of the pressure of the solar wind, the dayside boundary of the magnetosphere is located at a distance of about 10 earth's radii from the center of the earth, whereas the nightside boundary is at a distance of hundreds of earth's radii (as a comparison, the moon's orbit radius is of the order of 60 earth's radii). So, the nightside of the magnetosphere has the form of a long tail directed away from the sun.

The boundary between the earth's magnetic field and the solar wind is called the magnetopause. The solar wind is permanently changing, along with its pressure on the magnetopause. Energy, mass, and motion are transferred to the magnetosphere from the solar wind. Under the influence of these processes and the underlying ionosphere (a cold plasma with a concentration of neutral atoms, depending on the action of solar rays) and atmosphere, the plasmasphere, the ring current, and the radiation belts arise. The dynamic conditions of this magnetosphere–ionosphere plasma determine the complex of phenomena in the near-earth space, known as "space weather." Its impacts on the population (see Chaps. 5, 6, 7, and 8, devoted to the related neuro-physiological and cardio-physiological mechanisms) and technology are rather powerful (Liu and Fujimoto 2011; Campbell 2003).

4.3 The Ionosphere

The ionosphere is the domain of the upper earth atmosphere underlying the magnetosphere (see Sect. 4.2). The ionosphere is located at a height of 60–600 km above the earth and contains neutral molecules and atoms, along with ionized ones, mostly due to intensive collisions with solar ultraviolet and x-ray photons, and corpuscular radiation. In addition, cosmic rays play a part in producing ions and electrons in the ionosphere. Thus, it is a conductive medium. The ionization processes depend on the activity of the sun (e.g., the solar cycle period/sunspot cycle), the time (seasonal or daily changes), and geographical factors (polar, middle-latitude, and equatorial zones). The ionosphere includes the D, E, and F domains, which are usually referred to as layers; they have different properties and functions in the global ionosphere dynamics. The D layer is the lowest layer, and the boundary between it and the underlying nonconductive atmosphere domain may not be determined exactly. The D layer absorbs high-frequency radio waves, exists mainly during the daytime, and disappears gradually near nighttime. Amplitude modulation (AM) radio waves penetrate higher than the D layer and are reflected from there back to the earth, then the process of penetration into the ionosphere above the D layer is repeated. Higher, above the D layer, is the E layer (or the Kennely–Heaviside layer), which was discovered earlier than the other layers. This layer disappears at nighttime, weakening after sunset. The E layer is located at a height of 105 km and absorbs x-rays. The F layer (or Appleton layer) may be found above the E layer at a height of 150 km and has a higher concentration of charged particles than the other layers. Though its structure changes during the day, it is a relatively stable layer, absorbing extremes of ultraviolet radiation. The F layer includes the lower F1 layer and the F2 layer above F1, with greater electron density.

The free electrons in the ionosphere allow effective propagation of electromagnetic (EM) waves and radio communication. The ionosphere is the medium for the aurora.

There are other waves in the ionosphere, which are not described here (e.g., gravitational waves), and waves of different physical natures may interact. We shall not describe the composed nonlinear dynamics of the ionosphere plasma in more detail here, but its processes do have important influences on humans (see Chaps. 5, 6, 7, and 8) (Buchkov et al. 2014; Abdu et al. 2011).

4.4 Geomagnetic Storms and the Planetary Geomagnetic Index

Geomagnetic storms hinder the propagation of radio waves, navigation, electrical networks, and the Global Positioning System (GPS), and they accelerate deterioration of pipelines because of induction currents, etc. These technological processes and systems are controlled by human brains. The EM system of a brain is influenced by the EM fields of geomagnetic storms, as are technological systems.

According to the authors' research, a geomagnetic storm with a geomagnetic planetary index $K_p \geq 5$ at the latitude of Moscow considerably decreases the coherence of the brain's EM processes and changes cardiac parameters (see Chaps. 5, 6, 7, and 8, devoted to geo-biophysics).

Geomagnetic storms arise in the magnetosphere of the earth because of disturbances created there by variations in the solar wind's characteristics and resulting changes in currents, plasma parameters, and fields. These processes develop with considerable power during periods of relatively high velocity of the solar wind. A similar regime of magneto-hydrodynamic (MHD) flow is supported by the volume of solar energy received from the coronal mass ejection (CME) during the beginning stage of the solar wind flow formation. As one of the results of the solar wind process, huge masses of solar plasma (of the order of a billion tons) arrive in the near-earth cosmic space, along with solar magnetic field structures embedded in the mass of the CME plasma. The propagation of solar wind plasma from the sun to the earth typically takes several days, but in the case of the strongest ejections and the most rapid flow, the propagation takes only about 20 h.

There is another solar wind phenomenon known as the high-speed solar (HSS) wind stream. The characteristic feature of the HSS is its ability to interact with the slow solar wind stream and create co-rotating interaction regions (CIRs). Some moderate geomagnetic storms are created by participation of CIRs, and though the fields accompanying this process are not as strong as those in the case of other mechanisms of magnetic storm generation (e.g., CME), the integral period of action of CIRs may be long enough to deposit larger portions of energy in the earth's magnetosphere.

Changes in the radiation belts of the earth and heating of the ionosphere and thermosphere (the region of the upper atmosphere) arise as results of geomagnetic storms. Moreover, the westward ring of the electrical current around the earth in the near-earth space creates magnetic field changes near the earth and on its surface. As a measure of this current, the disturbance storm time (Dst) index, which is a characteristic of the magnetic storm size, has been used.

Though a number of considerable MHD phenomena (CME, HSS, CIR, changes in the radiation belts, and the electrical current around the earth) have been listed, there are another currents, produced in the magnetosphere, which follow the magnetic field there and are known as field-aligned currents (FACs, or Birkeland currents). These currents are essential for the linkage of the solar wind, i.e., a magnetosphere system, with intense currents in the auroral ionosphere. The auroral currents are known as auroral electrojets; these generate large magnetic disturbances on the ground, as well as other MHD phenomena that have been mentioned. All of these currents and magnetic disturbances on the earth's surface contribute to the magnitude of the planetary geomagnetic index, $K_{p.}$

This index is the basis used to derive the *G*-scale—one of the three National Oceanic and Atmospheric Administration (NOAA) space weather scales used to

Table 4.1 Conversion of magnetic disturbance data into the geomagnetic index (*K*-index) value

Magnetic disturbance (nT)	*K*-index value
0–5	0
5–10	1
10–20	2
20–40	3
40–70	4
70–120	5
120–200	6
200–330	7
330–500	8
>500	9

estimate space weather dynamics, sometimes resulting in interruption of the normal functioning of different systems on the earth, including humans and technology.

The K_p index is the weighted average value of the 3-h geomagnetic *K*-indices, which are calculated by 13 geomagnetic observatories located between northern or southern latitudes of 44° and 60°, with application of the recommendations of the International Association of Geomagnetism and Aeronomy (IAGA). Every *K*-index is calculated by the given observatory on the basis of the measured horizontal components of the three-component vector of magnetic induction, usually referred to as the magnetic field vector. The most disturbed horizontal magnetic component is chosen for the investigated 3-h time interval, and then the difference between the maximal and minimal magnitudes of this component for this time interval is calculated. The calculations are performed under the condition that the oscillation of the chosen component that is typical of a Q-day is removed (a Q-day denotes a magnetically quiet day; the day under investigation may be a D-day, i.e., a disturbed day). Special tables are used to convert the aforementioned magnetic difference into the geomagnetic *K*-index value (Table 4.1).

Table 4.1 shows the data used for the magnetometer in the Boulder SWPC in Colorado, USA. One can see that the increase in *K* values with increases in magnetic disturbance form a quasi-logarithmic scale. The maximal module of magnetic disturbance (positive or negative), needed to calculate the *K* values, may arise in every moment of the 3-h period [00:00–03:00, 03:00–06:00, . . . 21:00–24:00, according to universal time (UT)]. The maximal magnetic disturbances must not be missed, to ensure correct calculation of the *K* value for every 3-h period. This is the reason for 1-min monitoring of the geomagnetic field by the observatories preparing data for the official information on the geomagnetic field variations and prognosis. The latter is of importance not only for men and women with evaluated dependence on geomagnetic field changes, but also for healthy specialists controlling complicated technology, because geomagnetic storms are capable of interfering in the EM processes of the human brain (see Chaps. 5 and 6).

The necessity to apply the particular conversion table for every geomagnetic observatory is connected with the different geomagnetic conditions at the different latitudes. Indeed, the value $K = 9$ corresponds to a disturbance of more than 2500 nT in the region of the polar aurora and more than 300 nT only at low latitudes, except the near-equatorial regions (Bartels 1939; Gonzales et al. 1994).

4.5 Discussion

Let us recognize that the K_p index is sensitive to most geomagnetic phenomena, including ones that have been omitted from this short qualitative introduction to the theory of geomagnetism, e.g., EM field resonances in the space between the earth's surface and the ionosphere (Schumann resonance). Their influence has not been investigated by us from the viewpoints of electro-neurophysiology and/or (as in this chapter) other phenomena of space weather (the solar wind, dynamics of the ionosphere–magnetosphere plasma, geomagnetic storms, and the *K*-index). The *K*-index has been considered by us in connection with the geospheres and the phenomena above the earth that we have described here, but the earth's EM field, being generated in its depths, takes part in the formation of the *K*-index due to underground EM field dynamics, including terrestrial electrical currents and EM field precursors of earthquakes (see Chaps. 1, 2, and 3).

As can be seen from the above, biophysical factors are not included in any form in the algorithm for calculation of *K*-index and K_p index values, as a result of their averaging. Nevertheless, EM brain and cardiac processes are closely dependent on K_p, according to our measurements (see Chaps. 5, 6, 7, and 8). This fact may be explained by the supposition that these physiological processes depend on differences in geomagnetic disturbance magnitudes similar to the *K*-index.

References

Abdu, M. A., et al. (Eds.). (2011). *Aeronomy of the earth's atmosphere and ionosphere*. New York: Springer.

Bartels, J. (1939). The three-hour range index measuring geomagnetic activity. *Geophysical Research, 44*, 411–454.

Buchkov, V., et al. (Eds.). (2014). *The atmosphere and ionosphere*. New York: Springer.

Campbell, W. (2003). *Introduction to geomagnetic fields* (2nd ed.). Cambridge: Cambridge University Press.

Delamere, P., et al. (Eds.). (2015). *Solar wind interactions with giant magnetospheres and earth's magnetosphere, Geophysical monograph series*. Washington, DC: American Geophysical Union.

Gonzales, W. D., et al. (1994). What is a geomagnetic storm. *Journal of Geophysical Research, 99* (A4), 5771–5792.

Liu, W., & Fujimoto, M. (2011). The Dynamic Magnetosphere. Springer.

Meyer-Vernet, N. (Ed.). (2007). *Basics of solar wind, Cambridge atmospheric and space science series* (p. 478). Cambridge University Press: Cambridge.

Miralles, P. M., & Almeda, J. S. (2011). *The Sun, the solar wind, and the heliosphere*. New York: Springer.

Waltz, K. (2015). *Solar wind* (p. 216). New York: Callisto Reference.

Chapter 5
Structures of the Brain

Abstract In this chapter, we give a rather restricted volume of known information about the brain, in order to facilitate understanding of the results of our measurements devoted to the connection of the brain with the cosmos. We show in Chap. 6 that the brain's electromagnetic (EM) processes depend closely on space weather (see Chap. 4) and especially on geomagnetic storms. The mechanism of this dependence is not known (it is far from the only unknown mechanism of the electro-neurophysiology of the brain). We suppose that the changes we have measured in the distribution of electrical potentials on the surface of the scalp during geomagnetic storms result from the interaction of the brain's macro-EM processes with ones on the nerve cell (neuron) level. Thus, the structures of the brain and the neuron are discussed in this chapter.

An introduction to electroencephalography (EEG) is provided because EEG is the basis of our measurements. Moreover, a section on brain rhythms (i.e., the frequency ranges of the brain's EM processes) is included because, according to our EEG measurements, the theta rhythm is the phenomenon that most effectively demonstrates the distinctive dependence of the brain on space weather.

Keywords Electro-neurophysiology · Electroencephalography · Brain rhythms

5.1 The Problem

In this book on mathematical geo-biophysics, devoted to the interaction of geophysical fields and their impacts on the brain and the heart, it is important for the authors to describe the mechanism of this impact on the basis of the main physical principles and the proper mathematical models. In other words, a system of nonlinear partial differential equations (PDEs) (or maybe integro-differential equations, taking account of the complex properties of the medium of the brain) addressing the coupled chemical electro-thermo-viscoelasticity should be formulated along with the initial boundary value conditions.

O. Novik et al., *Electromagnetic Geophysical Fields*,
https://doi.org/10.1007/978-3-319-98461-2_5

In addition, the quantitative characteristics of the coefficients and the integral kernels of the system must be described, including the spatial and temporal distributions of these data and "forces," e.g., the electrical current density for the electromagnetic (EM) block of the model. Then we must investigate the mathematical correctness of the model and its discretization (algorithm). Now, calculations and visualizations of the EM signals on the surface of the scalp are needed, because such signals are observed in biophysical experiments, similar to the precursory signals on the surface of the earth.

However, we are not able to realize this program of brain research, though a proper program has been realized in the case of the geophysical problem, as described in the initial part of this book (namely, Chaps. 1, 2, and 3 with the mathematical background in the final Chaps. 9, 10, and 11).

Maybe the main unsolved problem of the theory of the brain's EM processes is the problem of an adequate schematic representation of the brain structures and their functioning in terms of electrodynamics and its interaction with the physical and chemical fields (fields of concentrations of chemicals) in the brain. The brain structures and their functions are far from being modeled correctly in frames of known physical, chemical, mathematical, and computational approaches.

Therefore, we shall consider the EM signals we have measured [with application of electroencephalography (EEG)] on the surface of the scalp without making an attempt to construct a theory of their origin on the basis of first principles (including neurophysiological data), though we have described a first principles' theory of the precursory EM signals of an earthquake on the land and marine surfaces of the earth.

EEG is a branch of neurology, recording, analyzing, and interpreting—from the point of view of neurological diseases—the distribution of electrical potentials and their dynamics on the surface of the scalp, without invasive methods.

We will apply an EEG approach for quantitative investigation of the influence of space weather dynamics, namely geomagnetic storms, on the brain. In other words, we will try to increase understanding of the following problem: What is the reaction of the EM system of the brain to EM extreme events in the ionosphere–magnetosphere plasma, and are some recommendations possible to attenuate any such negative influence?

This is of importance because plenty of complicated modern systems (e.g., nuclear power stations, rockets, aircraft, trains, and so on) must be under the operational control of a brain that is not disturbed by external EM fields, including ones generated in the ionosphere–magnetosphere plasma by geomagnetic storms. Let us recognize that technological and defensive systems have become more complex and the scale of sequences of control errors is increasing. These problems differ greatly from those arising from the usual medical applications of EEG, in which the particularities of an original recording or its transformation (using the fast Fourier transform (FFT), wavelets, and other methods) are used to recognize different neurological diseases, using accumulated experience of comparison of these particularities and diseases.

We have applied a coherence function (a known tool in the theory of signals and related domains) to the transformation of the EEG signals recorded by us, to recognize the influence of the dynamics of the ionosphere–magnetosphere plasma, namely geomagnetic storms, on the brain's EM processes. Below, we describe this investigation on the basis of some preliminary information about the brain's structures and functions, and on EEG. From our point of view, it is surprisingly that the brain's nerve cells (probably mostly neurons)—which, as a result of their integrative activity, form the cognitive processes of a brain—are able to be connected with the cosmos! But we shall demonstrate the influence of geomagnetic storms on the brain (see Chap. 6).

Maybe there is some participation of neuroglia—or, simply, glia—in the brain's connection with the cosmos. But we suppose that neurons play a leading role because the reaction to the cosmos is dependent on the functional load on the brain, according to the testing we have performed on subjects using a computer.

Certainly, there are other methods for investigation of the brain—magnetic resonance imaging (MRI), functional magnetic resonance imaging (fMRI), positron emission tomography (PET), and others—but they are all too expensive as compared with EEG. Besides, although they may have higher spatial resolution than EEG, the temporal resolution (up to tens of minutes) of some of them is poor, whereas the temporal resolution of an EEG is of the order of milliseconds. Thus, except for EEG, the aforementioned tools may hardly be acceptable to register the brain's reaction to dynamic processes such as magnetic storms. Possibly, magnetic encephalography (MEG) could be applied for these goals because, like EEG, it has good temporal resolution, but we have no experience with MEG.

Thus, we have applied the usual EEG measurements, from the point of view of the widespread technology of these processes.

With regard to the specifics of our application of EEG for investigation of the connection of a brain with the cosmos in a group of healthy young volunteers, it should be briefly noted here that the volunteers were not informed beforehand that their connection with the cosmos was being investigated (to avoid undesirable excitation of the examinees by the EEG recordings).

The authors hope to show researchers with different scientific experience—including those specializing in physical sciences, mathematical sciences, and information technology (IT)—that problems of electro-neurophysiology are interesting and are not unapproachable; e.g., applications of the methods of the theory of nonlinear dynamical systems and the theory of diffusion processes may be found in Izhikevich (2010), Ermentraut and Terman (2010), and Eckhoff et al. (2015).

Therefore, these problems may be considered as a domain of application of their efforts, and the mathematical and computational components of brain research will be made more effective. But, first of all, young physicists, mathematicians, and other representatives of the exact sciences will receive a short introduction for beginners to the structures and functions of the brain and heart, in a similar book with physical content (see Chaps. 1, 2, 3, 4) and mathematical content (see Chaps. 9, 10, 11).

The subjects we describe there are connected with these medical applications, but we omit them here because the goal of this chapter is to provide a minimal volume of known information about the brain in order to elucidate our findings about the influence of the cosmos on the brain (see Chap. 6). A similar problem regarding the heart is considered in Chaps. 7 and 8.

5.2 Structures and Functions of the Brain

The main brain structures, seen from outside the brain, are shown in Fig. 5.1.

We suppose that the macroscopic and cellular structures of the brain take part in creating its connection with the cosmos. Both structures should be described here; certainly, only a preliminary description is possible and rational here. On the other hand, the level of detail must be sufficient to allow understanding of our new findings regarding the influence of cosmic weather on the brain and heart.

Among the large variety of available figures showing the brain structures and functions, we have chosen illustrations from Dr. E. Bigler (of Brigham Young University; see http://www.waiting.com/brainfunction.html); Prof. P. Grobstain (of Bryn Mawr College; see http://serendip.brynmawr.edu/exchange/brains), who passed away on June 28, 2011; and P. A. Kinser (of Haverford College; see https://serendipstudio.org/exchange/brains/structures).

We believe that Figs. 5.1, 5.2, 5.3, 5.4, 5.5, 5.6, 5.7, 5.8, 5.9, 5.10, 5.11, and 5.12 (see also column 1 in Table 5.1), published by these scientists on the internet, are necessary for an interested nonspecialist to study for information about brain

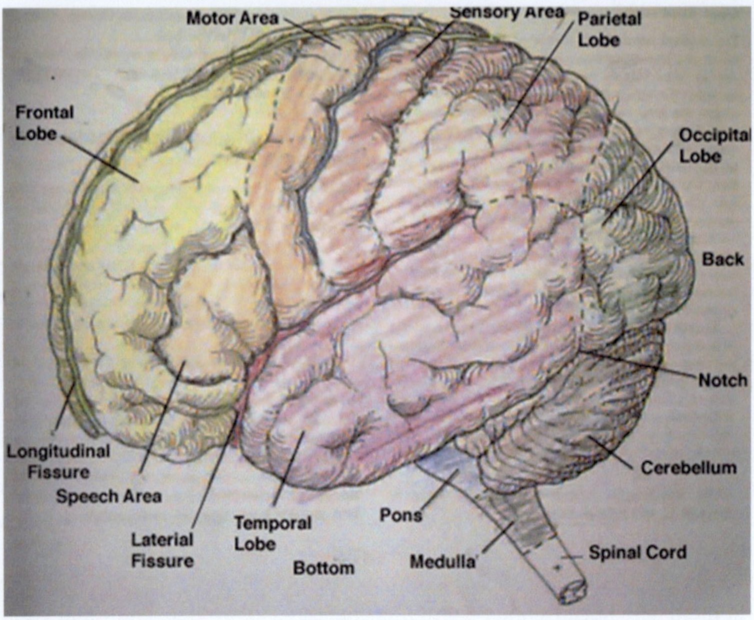

Fig. 5.1 The main brain structures. (Courtesy of Dr.E . Bigler, Prof. P. Grobstain, and P.A. Kinser, they published data in *open access* in Internet. This is related for all of the figures and Table devoted to the brain in this chapter)

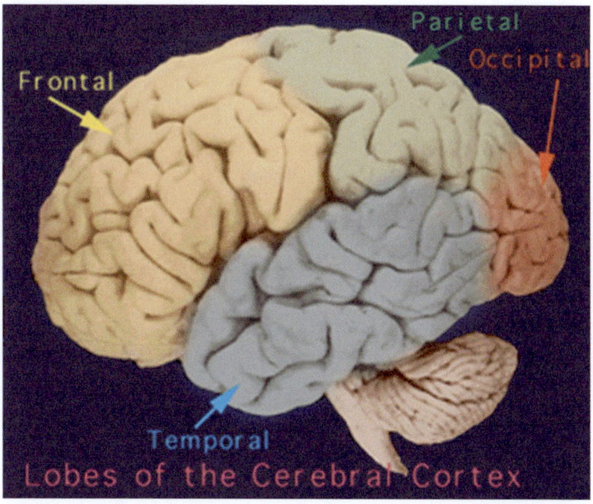

Fig. 5.2 Visual representation of the cortex

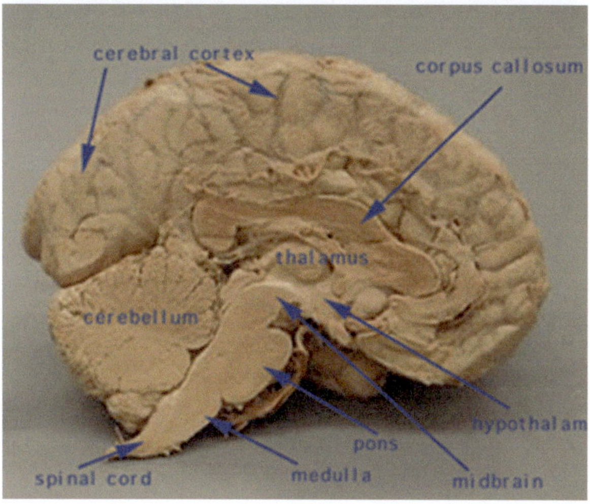

Fig. 5.3 Limbic system

structures and functions to understand our new findings, discussed in Chap. 6, about the connection of the brain with the cosmos. Moreover, the information given in this chapter regarding brain rhythms, neurons, EEG generation, etc., must be combined with the information on brain structures to consider the problem of the connection of the brain with the cosmos.

Now we will provide a short description of the central nervous system (CNS). It includes the brain, the spinal cord, and the peripheral nervous system (PNS). The CNS, with its nerves, reaches from the brain to the face, ears, eyes, nose, and spinal cord, and from the spinal cord to the rest of the body. Sensory nerves gather information from the environment and send it to the spinal cord, which then transfers the message to the brain. The brain then makes sense of that message and creates a

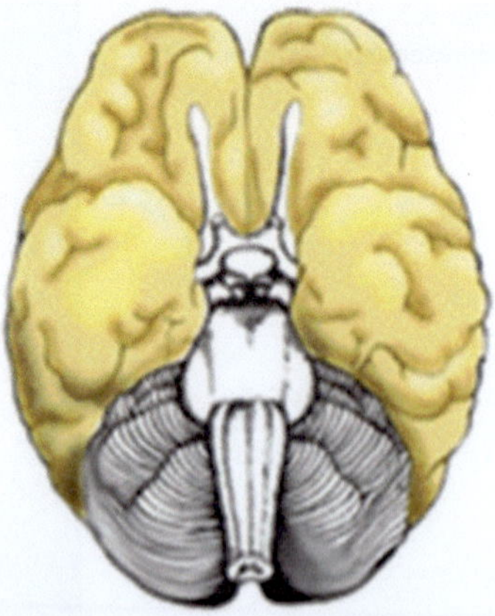

Fig. 5.4 Cerebral cortex: ventral view (from the bottom)

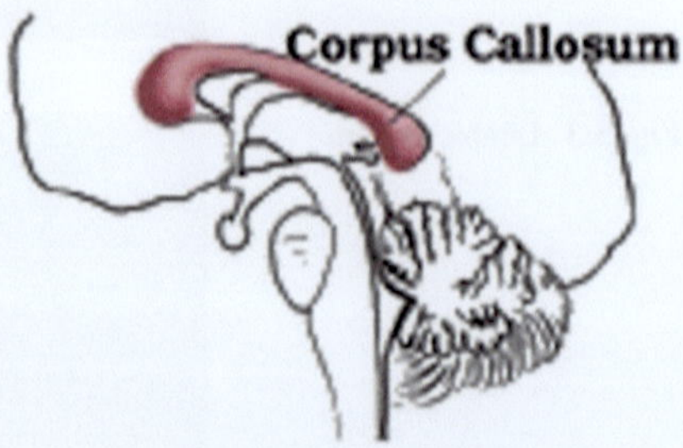

Fig. 5.5 Corpus callosum

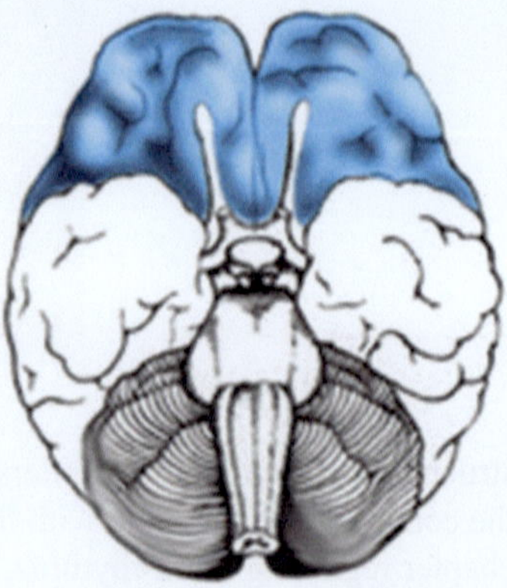

Fig. 5.6 Frontal lobe: ventral view (from the bottom)

response. Motor neurons deliver the instructions from the brain to the rest of the body. The spinal cord, made up of a bundle of nerves running up and down the spine, is similar to a superhighway, speeding messages to and from the brain every second. The brain is made up of three main parts: the forebrain, midbrain, and hindbrain. The forebrain consists of the cerebrum, thalamus, and hypothalamus (part of the limbic system). The midbrain consists of the tectum and tegmentum. The hindbrain is made

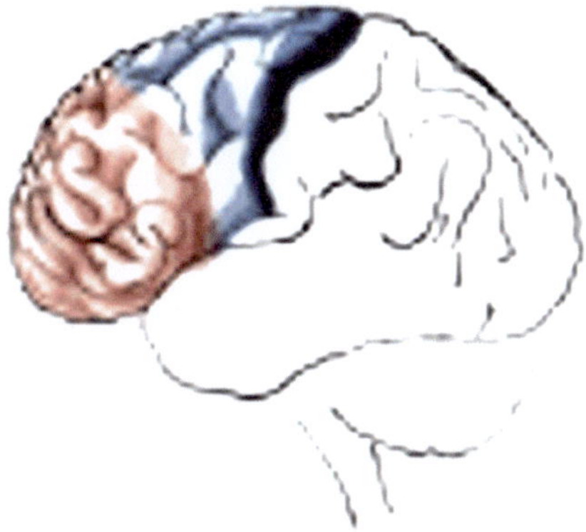

Fig. 5.7 Frontal lobe: side view

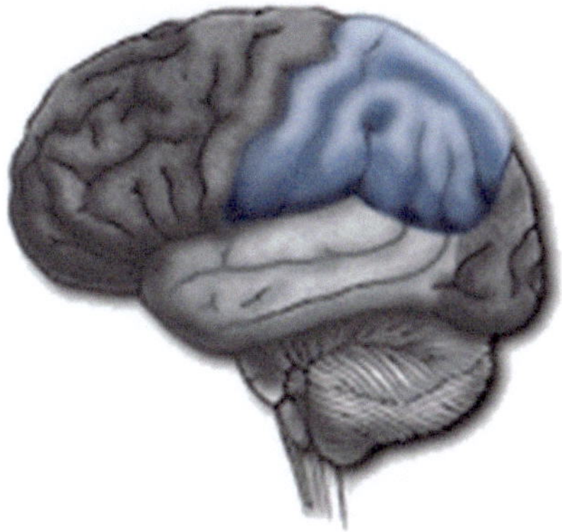

Fig. 5.8 Parietal lobe

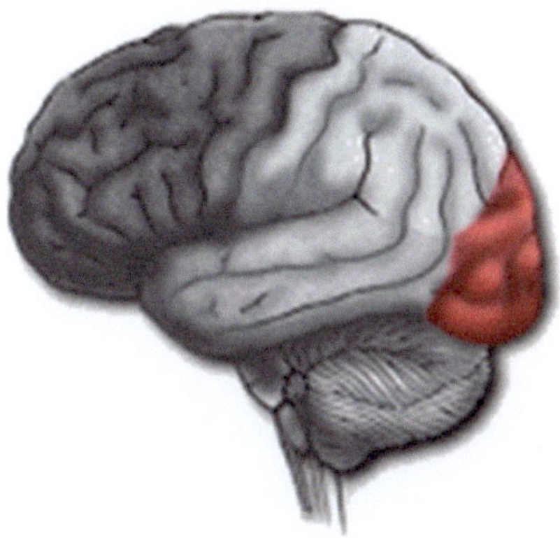

Fig. 5.9 Occipital lobe

up of the cerebellum, pons, and medulla. Often the midbrain, pons, and medulla are referred to together as the *brain stem*.

The cerebrum, or cortex, is the largest part of the human brain, associated with higher brain function such as thought and action. The cerebral cortex is divided into

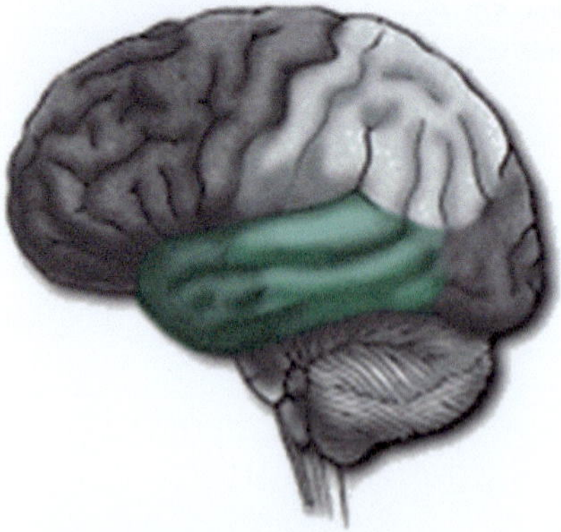

Fig. 5.10 Temporal lobe

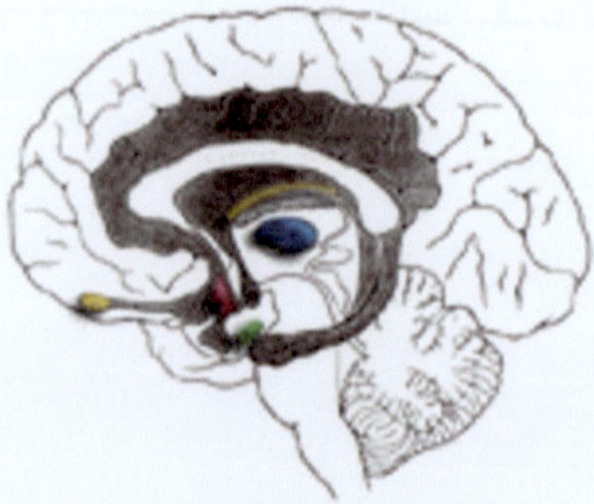

Fig. 5.11 Limbic system

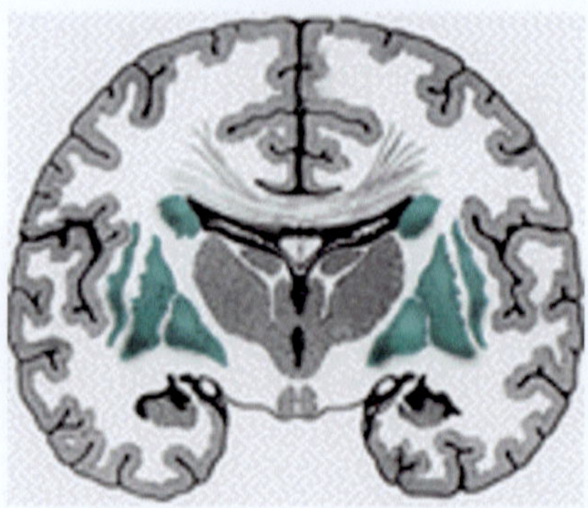

Fig. 5.12 Basal ganglia

four sections called “lobes”: the frontal lobe, parietal lobe, occipital lobe, and temporal lobe. Figure 5.2 provides a visual representation of the cortex.

- Frontal lobe: This is associated with reasoning, planning, parts of speech, movement, emotions, and problem solving.
- Parietal lobe: This is associated with movement, orientation, recognition, and perception of stimuli.

- Occipital lobe: This is associated with visual processing.
- Temporal lobe: This is associated with perception and recognition of auditory stimuli, memory, and speech.

Note that the human cerebral cortex is a great deal more wrinkled than the cerebral cortex in other animals' brains. Essentially, this makes the human brain more efficient because it increases the surface area of the brain and the number of neurons within it.

A deep furrow divides the cerebrum into two halves, known as the left and right hemispheres. The two hemispheres look mostly symmetric, yet it has been shown that each side functions slightly differently from the other. Sometimes the right hemisphere is associated with creativity and the left hemisphere is associated with logic abilities. The corpus callosum is a bundle of axons, which connects these two hemispheres.

Nerve cells make up the gray surface of the cerebrum, which is a little thicker than the thumb. White nerve fibers underneath carry signals between the nerve cells and other parts of the brain and body.

The neocortex occupies the bulk of the cerebrum. This is a six-layered structure of the cerebral cortex, which is only found in mammals. It is thought that the neocortex is a recently evolved structure and is associated with "higher" information processing by more fully evolved animals (e.g., humans, primates, and dolphins).

The cerebellum, or "little brain," is similar to the cerebrum in that it has two hemispheres and a highly folded surface or cortex. This structure is associated with regulation and coordination of movement, posture, and balance.

The limbic system, often referred to as the "emotional brain," is found buried within the cerebrum. Like the cerebellum, evolutionarily this structure is rather old. This system contains the thalamus, hypothalamus, amygdala, and hippocampus. Figure 5.3 provides a visual representation of this system, from a midsagittal view of the human brain.

We believe that the brief and distinctive information about the brain structures and functions given in Table 5.1, along with the aforementioned information, is useful before reading about the connection of the brain with cosmos, which is the main theme there.

The frontal lobe is located at the front of each cerebral hemisphere, positioned in front of the parietal lobes. The temporal lobes are located beneath and behind the frontal lobes. In the human brain, the precentral gyrus and the related cortical tissue that folds into the central sulcus comprise the primary motor cortex, which controls voluntary movements of specific body parts associated with areas of the gyrus. For further information about the frontal lobe and functions, see the relevant *Wikipedia* entry (Wikipedia Contributors 2018a).

Cognitive maturity associated with adulthood is marked by related maturation of cerebral fibers in the frontal lobes between the late teenage years and early adult years. The frontal lobe reaches full maturity around the age of 25 years. Research by Arthur Toga, from the University of California, Los Angeles (UCLA), found

Table 5.1 Brain structures and functions

Brain structure	Functions	Associated signs and symptoms
Cerebral cortex (see Fig. 5.4)	The outermost layer of the cerebral hemisphere, which is composed of gray matter. Cortices are asymmetric. Both hemispheres are able to analyze sensory data, perform memory functions, learn new information, form thoughts, and make decisions	
Left hemisphere	Sequential analysis: systematic, logical interpretation of information. Interpretation and production of symbolic information: language, mathematics, abstraction, and reasoning. Memory stored in a language format	
Right hemisphere	Holistic functioning: processing multisensory input simultaneously to provide "holistic" picture of one's environment. Visual spatial skills. Holistic functions such as dancing and gymnastics are coordinated by the right hemisphere. Memory is stored in auditory, visual, and spatial modalities	
Corpus callosum (see Fig. 5.5)	Connects right and left hemispheres to allow for communication between the hemispheres. Forms roof of the lateral and third ventricles	• Damage to the corpus callosum may result in "split brain" syndrome
Frontal lobe (see Figs. 5.6 and 5.7)	Cognition and memory Prefrontal area: the ability to concentrate and attend, elaboration of thought. The "gatekeeper" (judgment, inhibition). Personality and emotional traits Movement: motor cortex (Brodmann's): voluntary motor activity Premotor cortex: storage of motor patterns and voluntary activities Language: motor speech	• Impairment of recent memory, inattentiveness, inability to concentrate, behavior disorders, difficulty in learning new information. Lack of inhibition (inappropriate social and/or sexual behavior). Emotional lability. "Flat" affect • Contralateral plegia, paresis • Expressive/motor aphasia
Parietal lobe (see Fig. 5.8)	Processing of sensory input, sensory discrimination Body orientation Primary/secondary somatic area	• Inability to discriminate between sensory stimuli • Inability to locate and recognize parts of the body (neglect) • Severe injury: inability to recognize self • Disorientation of environment space • Inability to write
Occipital lobe (see Fig. 5.9)	Primary visual reception area Primary visual association area: allows for visual interpretation	• Primary visual cortex: loss of vision opposite field • Visual association cortex: loss of ability to recognize object seen in opposite field of vision, "flash of light," "stars"

(continued)

Table 5.1 (continued)

Brain structure	Functions	Associated signs and symptoms
Temporal lobe (see Fig. 5.10)	Auditory receptive area and association areas Expressed behavior Language: receptive speech Memory: information retrieval	• Hearing deficits • Agitation, irritability, childish behavior • Receptive/sensory aphasia
Limbic system (see Fig. 5.11)	Olfactory pathways: Amygdala and its different pathways Hippocampus and its different pathways Limbic lobes: sex, rage, fear; emotions. Integration of recent memory, biological rhythms Hypothalamus	• Loss of sense of smell • Agitation, loss of control of emotion. Loss of recent memory
Basal ganglia (see Fig. 5.12)	Subcortical gray matter nuclei. Processing link between thalamus and motor cortex. Initiation and direction of voluntary movement. Balance (inhibitory), postural reflexes Part of extrapyramidal system: regulation of automatic movement	• Movement disorders: chorea, tremors at rest and with initiation of movement, abnormal increase in muscle tone, difficulty initiating movement • Parkinson's

increased myelin in the frontal lobe white matter of young adults compared with that of teens, whereas the gray matter in the parietal and temporal lobes was more fully matured by the teenage years.

The so-called executive functions of the frontal lobes involve the ability to recognize future consequences resulting from current actions, choose between good and bad actions (or better and best), override and suppress unacceptable social responses, and determine similarities and differences between things or events.

The frontal lobes also play an important part in retention of longer-term memories that are not task-based. These are often memories with associated emotions, derived from input from the brain's limbic system, and modified by the higher frontal lobe centers to generally fit socially acceptable norms (see the preceding paragraph on executive functions). The frontal lobes have rich neuronal input both from the alert centers in the brain stem and from the limbic regions.

The human brain is the most complicated system known to us; there are doubts as to whether is it possible to understand the neurophysiology of the human brain by applying just the human brain! Certainly, it is necessary to increase our knowledge of it (see Figs. 5.13, 5.14, 5.15, and 5.16).

Now we turn from the macroscopic structures of the brain and their functions to the nerve cells. The mechanism of the influence of cosmic weather on the brain's functioning is unknown, but the fact that it exists is demonstrated in Chap. 6.

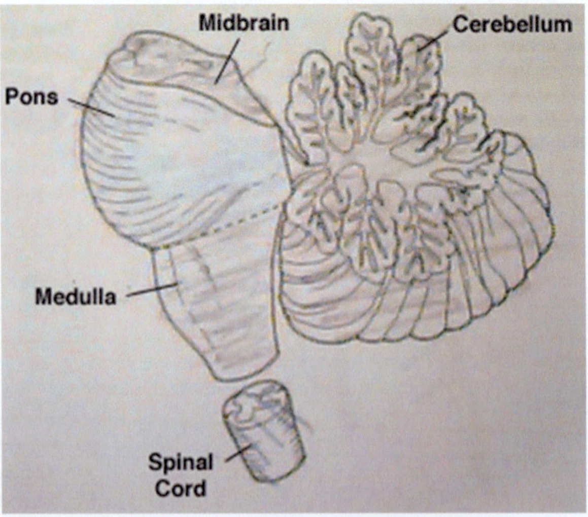

Fig. 5.13 Other brain structures

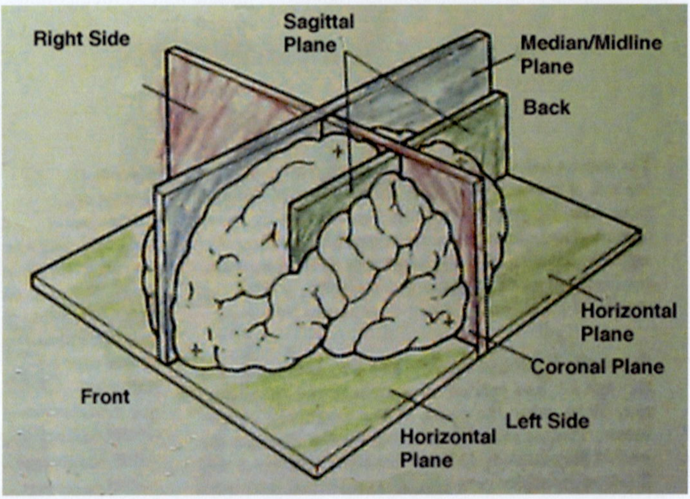

Fig. 5.14 Right and left sides of the brain

Now, we note only that all of the volunteers we examined demonstrated this influence clearly. We believe that the brain's nerve cells (neurons), with their ability to generate electrical currents, and the brain's macroscopic structures, with their ability to transport EM excitations in a measurable form to the surface of the scalp, are involved in the brain's reaction to geomagnetic storms (see below).

So, after brief consideration of the brain's structures on the macroscopic level, we turn our attention to the neurons.

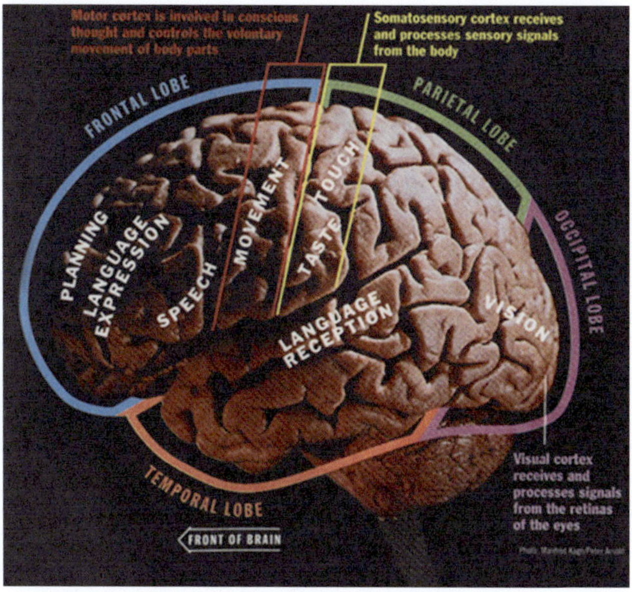

Fig. 5.15 Brain lobes and functions (note: these are discussed further in Chap. 6, which investigates the connection of the brain with the cosmos)

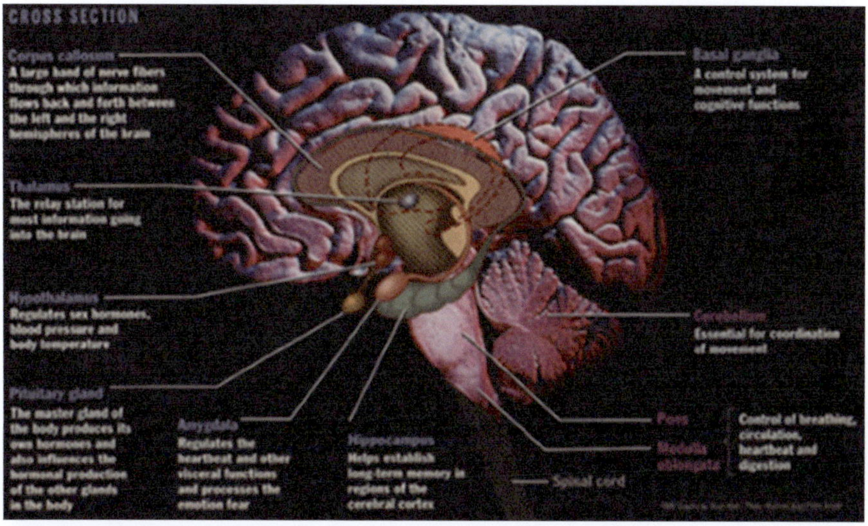

Fig. 5.16 Structures and functions of the brain (cross-section)

5.3 Neurons

In this section we describe other researchers' findings about neurons. The most fundamental and cited book on neurophysiology, in which the functioning of neurons, the EEG theory, and methods of recording brain signals are discussed, along with several other neurophysiological aspects, is *Niedermeyer's Electroencephalography: Basic Principals, Clinical Applications, and Related Fields* (Schomer and Lopes da Silva 2018).

The book *Basic Mechanism of the EEG*, from the Brain Dynamic Series (Zschocke and Speckmann 1993), is devoted to the difficult and important theme of the mechanisms of the EEG. In general, this theme has not been investigated well enough, considering the complex nature of the brain.

A short and clear introduction to EEG for beginners, related to geophysical processes and their impact on the brain and heart, sufficient to aid understanding of our book, may be found in *EEG Signal Processing*, written by Sanei and Chambers (2013). We follow them up to the end of the introduction to EEG for this reason and because their approach corresponds to the data given in most of the books about EEG that we know of. We remember that our research book was proposed to be accessible, among other applications, for lecture courses on mathematical geo-biophysics for postgraduates. Let us repeat: We have no new findings regarding neurons and the brain as a whole, except for *the connection of the brain with space weather dynamics* (in brief: *connection with the cosmos*).

So, we proceed by citing Sanei and Chambers (2013). Each nerve cell (neuron) consists of axons, dendrites, and cell bodies. A neuron responds to stimuli and transmits information. The body of a neuron has a single nucleus and contains most of the nerve cell metabolism, especially that related to protein synthesis. The proteins created in the cell body are delivered to other parts of the nerve. An axon is a cylinder, which transmits the form of the electrical impulses. In humans its length can range from a fraction of a millimeter to more than a meter.

Dendrites are connected to either the axons or dendrites of other cells and receive impulses from other nerves or relay signals to other nerves. In the human brain, each nerve is connected to more than 10,000 other nerves, mostly through dendritic connections [see the paper by Sanei and Chambers (2013)].

Following generation of an inhibitory postsynaptic potential (IPSP), there is an overflow of cations from the nerve cell or an inflow of anions into the nerve cell. This flow ultimately causes a change in potential along the nerve cell membrane. Primary transmembranous currents generate secondary inonal currents along the cell membranes in the intra- and extracellular spaces. The portion of these currents that flows through the extracellular space is directly responsible for the generation of field potentials. These field potentials, usually with less than 100 Hz frequency, are called EEGs when there are no changes in the signal average. The field potentials are by direct current (DC) if there are slow drifts in the average signals, which may mask the actual EEG signals (Fig. 5.17).

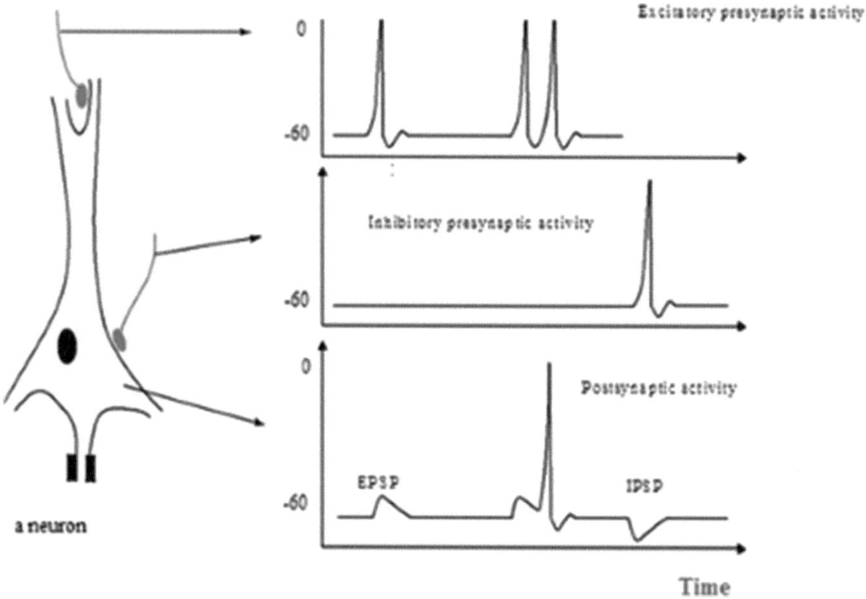

Fig. 5.17 The neuron membrane potential (measured in mV) changes and current flows during synaptic activation recorded by intracellular microelectrodes. Action potentials in the excitatory and inhibitory presynaptic fibers lead to excitatory postsynaptic potentials (*EPSPs*) and inhibitory postsynaptic potentials (*IPSPs*), respectively, in the postsynaptic neuron

5.4 Action Potential

Preliminary information about the action potential (AP) of a cell may be found in the relevant Wikipedia entry and the references cited there (Wikipedia Contributors 2018b). More detailed data on this basic function of a cell are given in a section titled "The Action Potential and Conduction of Electric Impulses" in Lodish et al. (2000).

APs are caused by an exchange of ions across the neuron membrane, and an AP is a temporary change in the membrane potential, which is transmitted along the axon. It is usually initiated in the cell body and normally travels in one direction. The membrane potential depolarizes (becomes more positive), producing a spike (Grun and Rotter 2010). After the peak of the spike, the membrane repolarizes (becomes more negative). The potential becomes more negative than the resting potential and then returns to normal. The APs of most nerves last between 5 ms and 10 ms.

The conduction velocity of APs lies between 1 m/s and 100 m/s. APs are initiated by many different types of stimuli; sensory nerves respond to many types of stimuli, such as chemicals, light, electricity, pressure, touch, and stretching. On the other hand, the nerves within the CNS (the brain and spinal cord) are mostly stimulated by chemical activity at synapses.

A stimulus must be above a certain threshold to set off an AP. Very weak stimuli cause a small local electrical disturbance but do not produce a transmitted AP. When the stimulus strength exceeds the threshold, an AP appears and travels down the nerve.

The spike of the AP is mainly caused by opening of sodium (Na^+) channels. The Na^+ pump produces gradients of both Na^+ and potassium (K^+) ions. Both are used to produce the AP; Na is high outside the cell and low inside it. Excitable cells have special Na^+ and K^+ channels with gates that open and close in response to the membrane voltage (voltage-gated channels). Opening the gates of Na^+ channels allows Na^+ to rush into the cell, carrying a positive charge. This makes the membrane potential positive (depolarization), producing the spike. For a human being, the amplitude of the AP ranges between approximately −60 mV and 100 mV. During this process, the following stages take place (see http://hyperphysics.phyastr.gsu.edu/hbase/biology/actpot.html):

1. When the dendrites of a nerve cell receive a stimulus, the Na^+ channels will open. If the opening is sufficient to drive the interior potential from −70 mV up to −55 mV, the process continues.
2. As soon as the action threshold is reached, additional Na^+ channels (sometimes called voltage-gated channels) open. The Na^+ influx drives the interior of the cell membrane up to approximately +30 mV. The process up to this point is called depolarization.
3. Then the Na^+ channels close and the K^+ channels open. Since the K^+ channels open much more slowly, the depolarization has time to be completed. Having both Na^+ and K^+ channels open at the same time would drive the system toward neutrality and prevent creation of the AP.
4. With the K^+ channels open, the membrane begins to repolarize back toward its resting potential.
5. The repolarization typically overshoots the resting potential to a level of approximately −90 mV. This is called hyperpolarization and would seem to be counterproductive, but it is actually important in the transmission of information. Hyperpolarization prevents the neuron from receiving another stimulus during this time, or at least it raises the threshold for any new stimulus. Part of the importance of hyperpolarization is in preventing any stimulus already sent up an axon from triggering another AP in the opposite direction. In other words, hyperpolarization ensures that the signal is proceeding in one direction.
6. After hyperpolarization, the Na^+/K^+ pumps eventually bring the membrane back to its resting state of −70 mV.

The nerve requires approximately 2 ms before another stimulus is presented. During this time no AP can be generated. This is called the refractory period.

5.5 EEG Generation

An EEG signal is represented by currents that flow during synaptic excitations of the dendrites of many pyramidal neurons in the cerebral cortex. When brain cells (neurons) are activated, synaptic currents are produced within the dendrites. These currents generate a magnetic field, measurable by electromyogram (EMG) machines, and a secondary electrical field over the scalp, measurable by EEG systems.

Differences in electrical potentials are caused by summed postsynaptic graded potentials from pyramidal cells that create electrical dipoles between the soma (body of a neuron) and apical dendrites, which branch from the neurons (Fig. 5.9). The current in the brain is generated mostly by pumping the positive ions of sodium (Na^+), potassium (K^+), and calcium (Ca^{++}), and the negative ion of chlorine (Cl^-), through the neuron membranes in the direction governed by the membrane potential (Atwood and MacKay 1989).

The human head consists of different layers, including the scalp, skull, brain, and many other thin layers in between them. The skull attenuates the signals approximately 100 times more than the soft tissue. On the other hand, most of the noise is generated either within the brain (internal noise) or over the scalp (system noise, or external noise). Therefore, only large populations of active neurons can generate enough potential to be recordable by scalp electrodes. These signals are later amplified greatly for display purposes (Fig. 5.18).

Approximately 10^{11} neurons are developed at birth when the CNS becomes complete and functional (Nunez and Srinivasan 2005; Shepherd 1991). This makes an average of 10^4 neurons per cubic millimeter. Neurons are interconnected into neural nets through synapses. Adults have approximately 5×10^{14} synapses. The number of synapses per neuron increases with age, whereas the number of neurons decreases with age. From an anatomical point of view, the brain may be divided into three parts: the cerebrum, cerebellum, and brain stem. The cerebrum consists of both left and right lobes of the brain, with highly convoluted surface layers called the cerebral cortex.

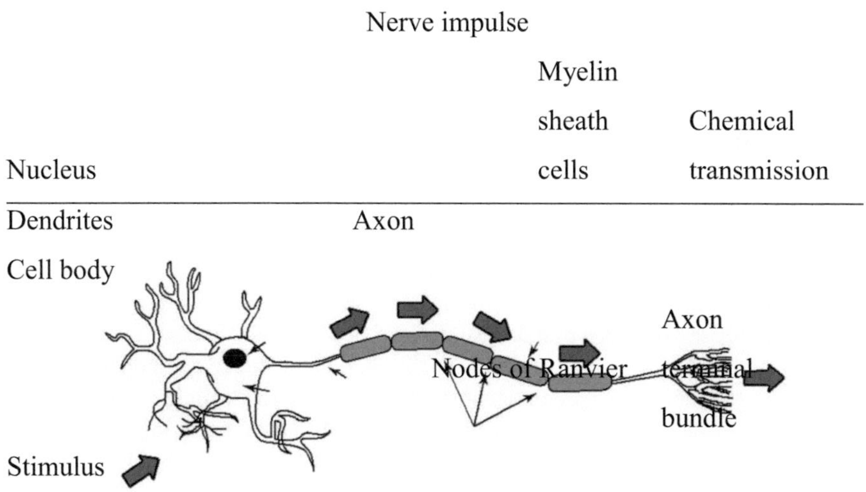

Fig. 5.18 Structure of a neuron. This figure is located in the public doman of Internet: https://www.khanacademy.org/science/biology/human-biology/neuron-nervous-system/a/overview-of-neuron-structure-and-function?utm_account=Grant&utm_campaignname=Grant_Science_Dynamic&gclid=EAIaIQobChMIqqzag8Pg3wIVCc-yCh0UIAcXEAAYAiAAEgInBPD_BwE

The cerebrum includes the regions for movement initiation, conscious awareness of sensation, complex analysis, and expression of emotions and behavior. The cerebellum coordinates voluntary movements of muscles and maintains balance. The brain stem controls involuntary functions such as respiration, heart regulation, biorhythms, and neurohormone and hormone sections.

5.6 Brain Rhythms

There are five major brain waves, distinguished by their different frequency ranges. These frequency bands, listed here from low to high frequencies, are called alpha (α), theta (θ), beta (β), delta (δ), and gamma (γ). The alpha and beta waves were described by Berger in 1929. Jasper and Andrews (1938) used the term *gamma* to refer to waves above 30 Hz. The delta rhythm was introduced by Walter (1938) to designate all frequencies below the alpha range. He also defined theta waves as those having frequencies within the range of 4–7.5 Hz. The notion of a theta wave was introduced by Walter and Dovey (1944).

Delta waves lie within the range of 0.5–4 Hz. These waves are primarily associated with deep sleep and may be present in the waking state. It is very easy to confuse artifact signals caused by the large muscles of the neck and jaw with a genuine delta response. This is because the muscles are near the surface of the skin and produce large signals, whereas the signal that is of interest originates from deep within the brain and is severely attenuated in passing through the skull. Nevertheless, by applying simple signal analysis methods to the EEG, it is easy to see when the response is caused by excessive movement.

Theta waves lie within the range of 4–7.5 Hz. The term *theta* might have been chosen to allude to its presumed thalamic origin. Theta waves appear as consciousness slips toward drowsiness. Theta waves have been associated with access to unconscious material, creative inspiration, and deep meditation. A theta wave is often accompanied by other frequencies and seems to be related to the level of arousal. It is known that healers and experienced mediators have an alpha wave that gradually lowers in frequency over long periods of time.

Alpha waves appear in the posterior half of the head and are usually found over the occipital region of the brain. They can be detected in all parts of the posterior lobes of the brain. For alpha waves the frequency lies within the range of 8–13 Hz and commonly appears as a round or sinusoid-shaped signal. However, in rare cases it may manifest itself as sharp waves. In such cases, the negative component appears to be sharp and the positive component appears to be rounded, similarly to the wave morphology of the Rolandic mu rhythm. Alpha waves have been thought to indicate both relaxed awareness without any attention, and concentration. The alpha wave is the most prominent rhythm in the whole realm of brain activity and possibly covers a greater range than has been previously assumed. A peak can regularly be seen in the beta wave range in frequencies even up to 20 Hz, which has the characteristics of an alpha wave state rather than those of a beta wave. Again, very often a response is seen at 75 Hz, which appears in an alpha setting. Most subjects produce some alpha waves with their eyes closed, which is why it has been claimed that it is nothing but a waiting or scanning pattern produced by the visual regions of the brain. It is reduced or eliminated by opening the eyes, by hearing unfamiliar sounds, by anxiety, or by mental concentration or attention. Albert Einstein could solve complex mathematical problems while remaining in the alpha state, although generally beta and theta waves are also present. An alpha wave has a higher amplitude over the occipital areas and has an amplitude of normally less than 50 μV. The origin and physiological significance of an alpha wave is still unknown, and so more research is needed to understand how this phenomenon originates from cortical cells.

A beta wave is the electrical activity of the brain varying within the range of 14–26 Hz (though in some literature, no upper bound is given). A beta wave is the usual waking rhythm of the brain associated with active thinking, active attention, focus on the outside world, or solving of concrete problems, and is found in normal adults. A high-level beta wave may be acquired when a human is in a panic state. Rhythmic beta activity is encountered chiefly over the frontal and central regions. Importantly, a central beta rhythm is related to the mu rhythm and can be blocked by motor activity or tactile stimulation. The amplitude of the beta rhythm is normally under 30 μV. Like the mu rhythm, the beta wave may be enhanced by a bone defect and also in tumoral regions.

Frequencies above 30 Hz (mainly up to 45 Hz) correspond to the gamma range (sometimes called the *fast beta wave*). Although the amplitudes of these rhythms are very low and their occurrence is rare, detection of these rhythms can be used for confirmation of certain brain diseases. The regions of high EEG frequencies and the highest levels of cerebral blood flow (as well as oxygen and glucose uptake) are located in the frontocentral area. The gamma wave band has also been shown to provide a good indication of event-related synchronization (ERS) of the brain and can be used to demonstrate the locus for right and left index finger movement, right toes, and the rather broad and bilateral area for tongue movement.

Figure 5.19 shows typical normal brain rhythms with their usual amplitude levels (Buzsaki 2006).

The delta wave is observed in infants and sleeping adults, and the theta wave is observed in children and sleeping adults. The alpha wave is detected in the occipital brain region when there is no attention, and the beta wave appears frontally and parietally with low amplitude.

On the scalp, the amplitudes commonly lie within the range of 10–100 μV.

The above rhythms may last if the state of the subject does not change, and therefore they are approximately cyclic in nature.

Other brain waveforms have not been described here, because the goal of this book's authors is to cite the known works on the brain up to the minimal level sufficient for beginners to understand our quantitative findings about the connection of the brain with the cosmos.

Let us note that in our investigations (see Chap. 6), the theta rhythm exhibited the most distinctive reaction of the brain to geomagnetic storms.

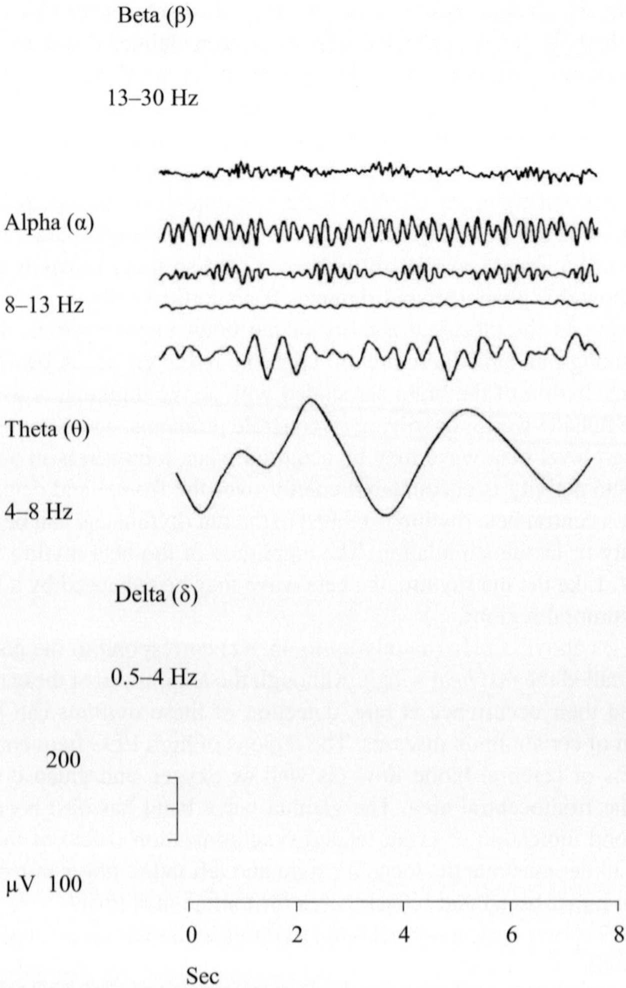

Fig. 5.19 Four typical dominant brain normal rhythms, from high to low frequencies

References

Atwood, H., & MacKay, W. (1989). *Essentials of neurophysiology*. Toronto: Decker.

Berger, H. (1929). Über das Elektrenkephalogramm des Menschen. *Archiv für Psychiatrie und Nervenkrankheiten, 87*(1), 527–570.

Buzsaki, G. (2006). *Rhythms of the brain*. Oxford: Oxford University Press.

Eckhoff, P., Holmes, P., et al. (2015). *A short course in mathematical neuroscience. Program in applied and computational mathematics*. Princeton: Princeton University.

Ermentraut, G. B., & Terman, D. H. (2010). *Mathematical foundations of neuroscience*. New York: Springer.

Grun, S., & Rotter, S. (Eds.). (2010). *Analysis of parallel spike trains*. New York: Springer.

Herbert H. J. (1938). Electro-Encephalography. *Archives of Neurology & Psychiatry, 39*(1), 96.
Izhikevich, E. (2010). *Dynamical systems in neuroscience: The geometry of excitability and bursting* (p. 441). Cambridge: MIT Press.
Jasper, H., Andrews, H. (1938). Normal differentiation of occipital and precentral regions. *Archive of Neurology and Psychiatry, 39*(1), 96–115. doi:10.1001/archneurpsyc.1938.02270010106010. Accessed 08 Jan 2019.
Lodish, H., Berk, A., et al. (2000). *Molecular cell biology* (4th ed.). New York: Freeman.
Nunez, P., & Srinivasan, R. (2005). *Electrical fields of the brain: Neurophysics of EEG* (p. 640). Oxford: Oxford University Press.
Sanei, S., & Chambers, J. (2013). *EEG signal processing*. Hoboken: Wiley.
Schomer, D.L., & Lopes da Silva, F. (2011). Niedermeyer's Electroencephalography: Basic Principles, Clinical Applications, and Related Fields (6 edition). Oxford University Press.
Schomer, D., & Lopes da Silva, F. (Eds.). (2018). *Niedermeyer's electroencephalography: Basic principles, clinical applications, and related fields.* (7th ed.p. 1264). Oxford: Oxford University Press.
Shepherd, G. M. (1991). *Foundations of the neuron doctrine*. Oxford: Oxford University Press.
Walter, W. G. (1938). Critical review: the technique and application of electro-encephalography. *Journal of Neurology, Neurosurgery & Psychiatry, 1*(4), 359–385.
Walter, W.G. & Dovey, V.J. (1944). Electro-encephalography in cases of subcortical tumour. *Journal of Neurology, Neurosurgery & Psychiatry, 7*(3–4), 57–65.
Wikipedia Contributors. (2018a). *Frontal lobe. Wikipedia, the free Encyclopedia.* https://en.wikipedia.org/w/index.php?title=Frontal_lobe&oldid=864583753
Wikipedia Contributors. (2018b). *Action potential. Wikipedia, the free Encyclopedia.* https://en.wikipedia.org/w/index.php?title=Action_potential&oldid=863294233
Zschocke, S., & Speckmann, J. (Eds.). (1993). *Basic mechanisms of the EEG*. Boston: Birkauser.

Chapter 6
Influence of Space Weather on the Bioelectrical Activity of the Human Brain

Abstract A study was performed to investigate the influence of geomagnetic storms at the latitude of Moscow on the electrical activity of the human cerebral cortex. It was shown by electroencephalographic (EEG) measurements that when healthy volunteers aged 18–23 years were performing tasks on a computer during a moderate geomagnetic storm or within 24 h after it, the coherence function values for electrical oscillations in the frontal polar and occipital areas of their brains in the 4.0- to 7.9-Hz range (theta rhythm) decreased typically by a factor of two or more (relative to the values during quiet geomagnetic conditions), although the arterial blood pressure, respiratory rate, and electrocardiographic features registered in parallel did not deviate from their norms.

Keywords Electro-neurophysiological effect · Geomagnetic storms · EEG time series

6.1 Introduction

In recent decades there has been ever-increasing scientific interest in the influences of solar/terrestrial physical processes on the earth's biosphere and human beings. It is not just studies on a qualitative level that have been vigorously pursued to describe such influences on various human functional systems, but also research applying physical digital measurements and mathematical data processing for numerical characterization of cosmo-geophysical influences on humans (Haken 1996; Aamodt and Wang 2008).

In this work we formulate a task for detecting any influence of magnetic storms (MSs) on the electrical activity of the brain in healthy young volunteers living at middle latitudes—or conversely, demonstrating the lack of such influence—using only well-proven (i.e., standard in the field, though perhaps not the most sensitive) means of measurement and plain processing of the primary digital data. Excluding (at the first step) any methodical novelties, we thereby become able to achieve an unambiguous result: If an influence was detected, this was so because it existed in

O. Novik et al., *Electromagnetic Geophysical Fields*,
https://doi.org/10.1007/978-3-319-98461-2_6

terms of the generally accepted methodology; whereas if such novelties were applied initially, there would be uncertainty related to the possible distorting effects of insufficiently proven measurement and computing methods. In the authors' opinion, more advanced means would need to be applied at the stage of refinement and extension of the results if some influence was detected, or in an attempt to detect it if the standard methods had failed.

Navigation, communication, transport, electrical energy distribution, and other technological processes and systems (gas and oil pipelines) are damaged by geomagnetic storms. It is hardly possible to suppose that the electromagnetic (EM) processes of a human brain are independent of powerful EM disturbances of the magnetosphere–ionosphere plasma, including geomagnetic storms, caused by solar activity. But technologies, with their increasing completeness, scale, and economic and social importance, are controlled by humans, and their functional systems are not protected from helio-geophysical dynamics (Breus and Rappoport 2003; Vladymyrsky et al. 2004; Chyzhevsky 1995; Novik and Smirnov 2007, 2008a, b, 2015; Smirnov 2007).

We aimed to establish whether it was possible to numerically quantify the characteristics of the influence of a moderate geomagnetic storm (these events are more frequent than strong geomagnetic storms) on the human brain, using standard measurements of electroencephalograms (EEGs) and mathematical processing of the recorded time series of oscillations of electrical potentials in the cerebral cortex in healthy young volunteers (students), in the normal conditions of their lives, i.e., without hospital isolation and without possible changes in their psycho-physiological state due to unusual (from their viewpoint) testing (see below). We decided that use of standard testing methods would be preferable at this stage because their properties, including artifacts, are better known.

A Holter-monitoring approach to electrocardiographic (ECG) measurements was used by S. V. Pobachenko and A. G. Kolesnik in the case of EEG measurements to establish the influence of helio-geophysical conditions, including the influence of Schumann resonance, on the dynamics of amplitude and frequency parameters of EEGs in volunteers aged 20–55 years without expressed pathologies (Pobachenko and Kolesnik 2012). Whereas those authors compared the characteristics of the geophysical EM field and those of the brain (Merzlyy et al. 2012), we investigated changes in correlations between EEG time series caused by moderate geomagnetic storms.

We use the term "background" recordings (or EEGs, cartograms, data, etc.) for measurements made without the influence of an MS, i.e., with no such storm occurring on the day of the measurements nor on the two preceding days. We use the term "storm" recordings (or EEGs, cartograms, data, etc.) in the opposite case.

In a study reported by Babaev and Allahverdiyeva (2005), performed in a group of healthy adult women in a state of relaxation, in a state of photo-stimulation, and in the process of hyperventilation (in the environment of an isolated room or hospital), qualitative visible differences between background EEGs, cartograms, and blood characteristics, on the one hand, and the same parameters under the influence of a major geomagnetic storm, on the other hand, were demonstrated. Medical estimations of the changes under the influence of geomagnetic storms were formulated.

We were attempting to answer the question about a moderate MS's influence on the human brain (see above) in the following quantitative sense. We constructed a function with numerical values, determined in sets of EEG recordings. If the values of this function differed between the set of the background recordings and the set of the storm recordings, we concluded that there is an influence of geomagnetic storms on the brain. The intensity of the influence could be characterized as a difference in the value of the constructed function between the set of storm recordings and the set of background recordings.

6.2 A Storm Influence on Human Brain EM Processes Was Not Discovered Without Frequency Filtering or With Its Application

We begin here with some examples of the EEG recordings used for further analysis.

Initially, the EEG recordings were performed in the absence of an MS and then in the presence of a moderate MS (see Figs. 6.1 and 6.2); processing was not applied in either of these cases. We failed to find a clear difference between the records in the presence of a MS and those in the absence of one.

Then, to discover the influence of an MS on the brain's bioelectrical processes, we filtered the original records in the frequency range of 4 ÷ 7.9 Hz, i.e., the electroneurophysiological theta rhythm, known well in electroencephalography, was considered (Figs. 6.3 and 6.4). Once more, no difference between the records obtained in quiet conditions and those obtained in disturbed geomagnetic conditions was registered by us, though the above filter had been applied.

Let us note that the expressions "without the influence of an MS" and "in the absence of an MS" are applied here for the case of $K_p < 5$.

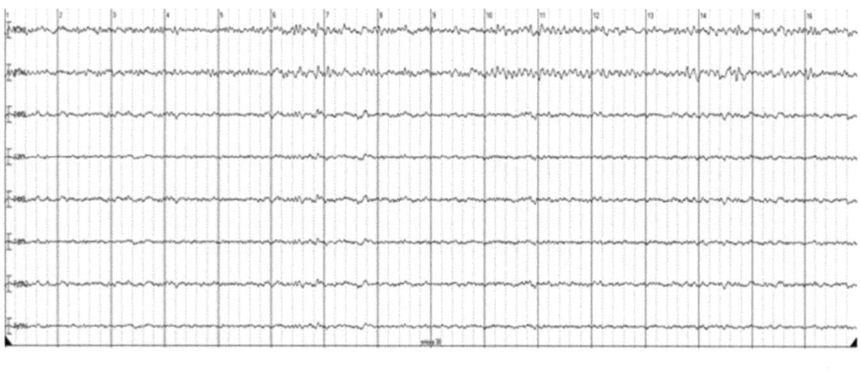

Fig. 6.1 Encephalographic (*EEG*) records (with 8 channels) in quiet geomagnetic conditions (geomagnetic index (K_p) = 2)

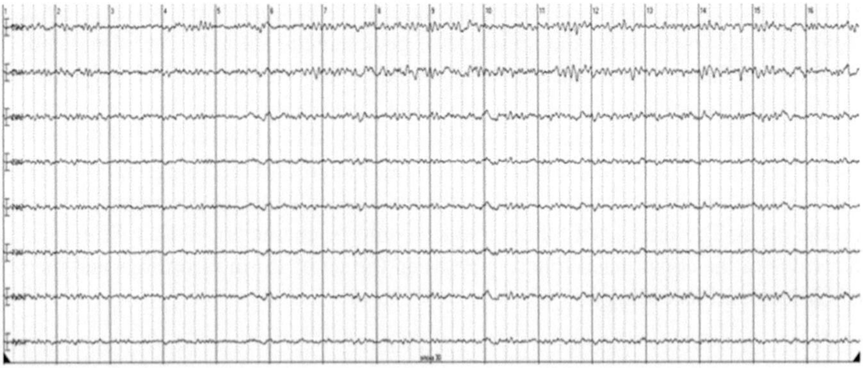

Fig. 6.2 Encephalographic (*EEG*) records (with 8 channels) under the influence of a geomagnetic storm (geomagnetic index (K_p) = 5)

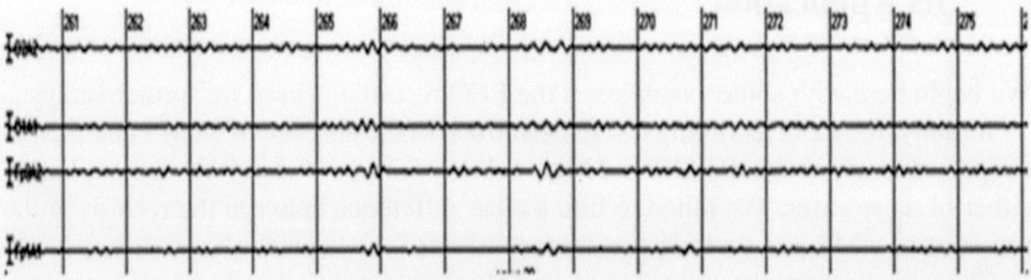

Fig. 6.3 Time series of theta rhythm oscillations (amplitude (*Ampl.*) = 20 μV, 4 ≤ frequency (*freq.*) ≤ 7.9 Hz) of electrical potentials in the cerebral cortex during functional loading (a proof-reading test) on a brain in quiet geomagnetic conditions; a filter in the frequency range of the theta rhythm was used

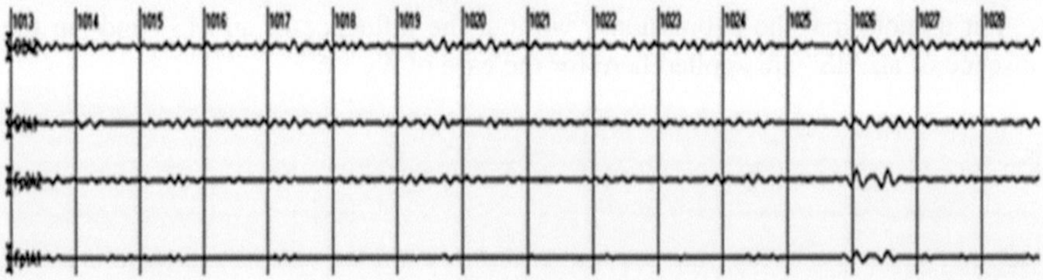

Fig. 6.4 Time series of theta rhythm oscillations (amplitude (*Ampl.*) = 20 μV, 4 ≤ frequency (*freq.*) ≤ 7.9 Hz) of electrical potentials in the cerebral cortex during functional loading (a proof-reading test) on a brain under the influence of a geomagnetic storm (geomagnetic index (K_p) = 5); the electroencephalographic recording was performed just during the storm

So, according to Figs 6.1, 6.2, 6.3, and 6.4, we could not differentiate between the EEG recordings in the absence of an MS and those under the influence of an MS, even in our attempts with frequency filtering.

Possibly, to determine the influence of an MS on the EM processes of the brain, a more effective way to process EEG records is needed.

6.3 Experimental Approach

The electro-neurophysiological effect of MSs is described by us for a group of 15 healthy students (volunteers) of both sexes, aged from 18 to 23 years. Their EEGs were registered from September 2004 to June 2005 on random days when the examinees had time for an EEG session, without scheduling the event to the course or forecast of geomagnetic situation. Monitoring of geomagnetic conditions was performed with Pushkov Institute of Terrestrial Magnetism, Ionosphere and Radio Wave Propagation (IZMIRAN) equipment, but neither the current geomagnetic state nor a geomagnetic forecast was made known to the examinees, to avoid psychological "attuning to MS" (as has been observed in elderly persons who have repeatedly experienced worsening of health associated with MS). In each EEG session, the electrical potentials in the brain hemisphere cortex were recorded for 2 min with the examinee's eyes open and with their eyes shut. Within each session, the examinees were also subjected to a known "proofreading test" (PRT; see below) on a computer, which took 15–30 min. During EEG recording, the examinees underwent control measurements of their blood pressure and respiratory rate, and an ECG; these characteristics did not deviate from the normal ranges in any examinee.

EEGs were recorded by a standard technique with a standard eight-channel electroencephalograph from Neiron–Spektr (Russia). Let us note that an EEG device with a large number of channels (typically, up to a few tens or more) is not an optimal tool for investigation of the influence of space weather on the brain's EM processes, because of the possibility of attenuation of the penetration of the EM field of a geomagnetic storm into the skull.

The left part of the electroencephalograph, with four active electrodes, registered the electrical potentials (relative to the "passive" electrode placed on the left earlobe) from four points at the left side of the head: on the forehead (frontal pole), on the top of the head (sinciput), on the temple, and on the back of the head (occiput). The right-side group of four electrodes registered the electrical potentials (relative to the right earlobe) at symmetric points of the right hemisphere. Pickups of 0.01 μV sensitivity took 256 readings of electrical potential at each point per second of EEG recording. The data were stored on the hard disk of a computer.

In the PRT, each examinee was presented consecutively with 40 lines of random symbols on a computer display and given the task of picking out a specified combination of symbols. The examinees were told that they had unlimited time to complete this task (usually, 15–30 min was sufficient). The computer registered the number of correctly and erroneously determined symbol combinations, but the examinees were not informed during the testing how successful their performance was. In this way, we excluded additional stress caused by thoughts such as "I can't do it in time" or "I am making too many annoying mistakes."

The results of the EEG measurement (or indices derived therefrom) were examined in two ways.

In the first approach, we compared the time series recorded during/after an MS and in quiet conditions (background) for every examinee; this could be considered a "personal" approach, and the series analyzed were pairs of MS versus background EEG records of the same examinee.

In the other approach, we considered two sets of time series: set A of background records and set B of records under the influence of an MS, without regard to a particular examinee. Thus, to answer the question of the effect of the MS on brain electrical activity, we needed to look for distinctions between sets B and A (other than the number of elements); this could be considered a "sets" approach. Note that in this way a larger number of EEGs could be used for analysis, including unpaired records (say, only background or only MS-associated EEGs from some examinees), in contrast to the personal approach, which required both background and MS EEG records for each examinee.

A distinct reaction of the brain electrical potentials to an MS was obtained by us in terms of the coherence function (CF), which is well known in mathematical statistics and in EEG analysis; it is determined with normalization of the cross-spectral density of two processes by their partial spectral densities. The CF value (CFV) defines the extent of coordination of the processes, with values in the interval [0, 1]; if one process is a transform of the other, the CFV equals unity.

Thus, here we assessed the influence of an MS on the coherence of electrical oscillatory processes in several pairs of points over either brain hemisphere.

From Figs. 6.5 and 6.6, one can see the decrease in the number of cases of low CFVs with an increase in the time interval after the end of the MS. Similar dynamics may be interpreted as a return to the coherence typical for the background geomagnetic conditions.

In Fig. 6.7, one can see that the number of measurements was increased considerably. Nevertheless, there were only three cases of high 1/CFV, namely up to 10, in contrast to the cases of measurements under the influence of an MS (Fig. 6.8).

6.4 Results and Discussion Regarding Rhythms of Brain Bioelectrical Oscillations

Analysis of EEG records usually includes isolation of the following five frequency ranges (also referred to as *rhythms*) typical for bioelectrical activity of the brain: delta (frequency $f = 0.5$–3.9 Hz), denoted as D; theta ($f = 4.0$–7.9 Hz), denoted as T; alpha ($f = 8.0$–13.0 Hz), denoted as A; beta low ($f = 14.0$–19.9 Hz), denoted as BL; and beta high ($f = 20.0$–35.0 Hz), denoted as BH.

On an EEG, the leads are denoted with letters specifying the position of the active electrode (e.g., Fp for frontal pole, O for occipital); the left-side electrodes are marked with "1" and the right-side electrodes are marked with "2." Thus, to correlate

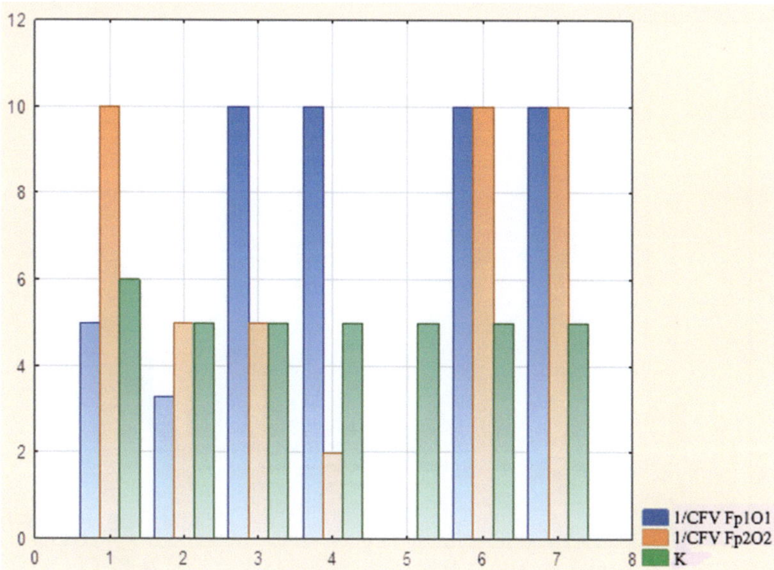

Fig. 6.5 Dependence of 1/*CFV* (coherence function value, shown on the *vertical axis*) on the *K*-index in the case of measurements during a magnetic storm. The number of measurements is shown on the *horizontal axis*

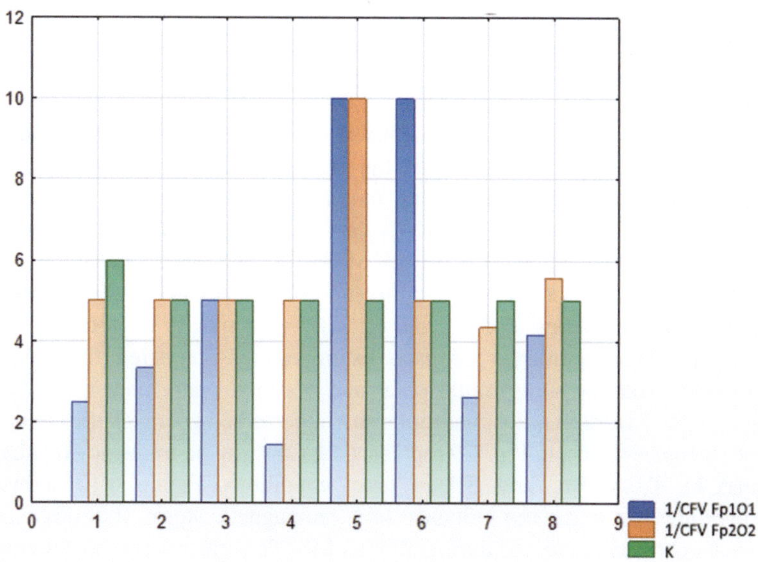

Fig. 6.6 Dependence of 1/*CFV* (coherence function value, shown on the *vertical axis*) on the *K*-index in the case of measurements 24 h after the end of a magnetic storm. The number of measurements is shown on the *horizontal axis*

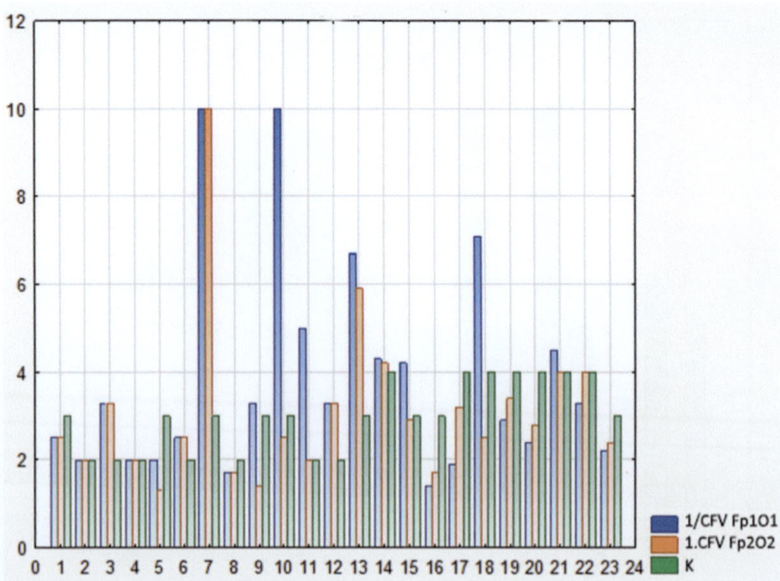

Fig. 6.7 Dependence of 1/*CFV* (coherence function value, shown on the *vertical axis*) on the *K*-index in the case of measurements during quiet geomagnetic conditions. The number of measurements is shown on the *horizontal axis*

the frontal polar and occipital series recorded from the left hemisphere, the resultant notation is "Fp1O1" and "Fp2O2" is the corresponding notation for the right hemisphere (see figures). We calculated the CFVs for oscillations in each brain rhythm, and these are shown as the heights of the respective bars in the figures.

In the notation for particular records, the first letters (Ka, Gr, . . . Kl) specify the examinee, followed by the date (in day–month–year order) and the maximal geomagnetic *K*-index on that day (IZMIRAN data: *K*3 or less denotes quiet conditions and *K*5 or more denotes an MS). In the MS-associated cases, the numbers after *t* denote the number of hours between the end of the MS and the start of the EEG recording (*t*0 means during an MS), and the last letters specify the test.

In the personal approach, for each examinee we compared the coherence between—for example—the frontal polar and occipital brain lobes (see Chap. 5) during computer work in quiet conditions and in perturbed geomagnetic conditions. The six examples shown in Fig. 6.9 represent the CFVs in the frequency ranges (i.e., rhythms) D, T, A, BL, and BH in quiet conditions. Figure 6.10 shows the corresponding data under the influence of a geomagnetic storm. It is obvious that coherence generally decreased during/after an MS. The effect was most pronounced and most regularly observed in the slower rhythms (D and especially T); in all pairwise comparisons, when the T coherence was considerable in quiet weather, it dropped by at least half in the presence of an MS. For the faster rhythms, the effects were smaller and less consistent; quite often the BH coherence did not decline at all.

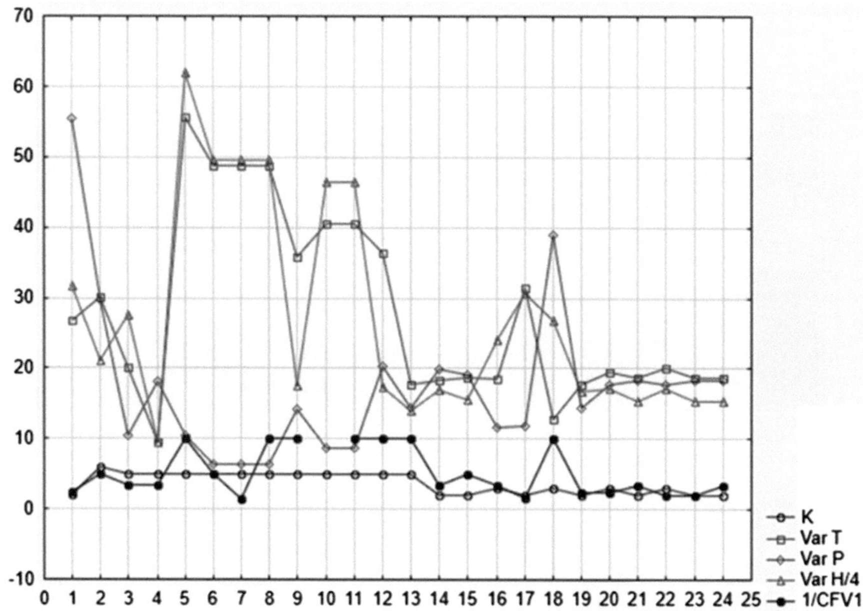

Fig. 6.8 Variations in temperature (*Var T*, measured in °C and denoted by *squares*), pressure (*Var P*, measured in mm Hg (torr) and denoted by *diamonds*), humidity (1/4 *Var H*, measured in % and denoted by *triangles*); and dynamics of the *K*-index (denoted by *white circles*) and bioelectrical activity (*1/CFV1*, denoted by *black circles*). The *horizontal axis* shows the number of measurements. The *left vertical axis* shows the temperature, pressure, and humidity variations, which had no influence on bioelectrical activity. However, one can see the correspondence of the *K*-index and bioelectrical activity (1/CFV1). *CFV* coherence function value

Note that CFV drops were evident even 24 h after an MS (see cases Gr and Kl), but no such effect was observed at 48 h (not shown).

It should be admitted that without the PRT, eyes-shut or eyes-open EEGs taken in the same sessions did not reveal any appreciable MS-associated change in Fp1O1 or Fp2O2 coherence; likewise, no distinct effect of an MS on CFVs was observed for any other combination of EEG leads (with PRT or without), so those data are not displayed here. It should also be borne in mind that variations in EEG responses to an MS may be associated not only with individual features of examinees (e.g., asymmetry in Gr, or a drop to zero in Mat) but also with some characteristics of a particular MS that are not reflected in the *K*-index but exert specific effects on brain activity.

Aiming to answer the question about the influence of an MS on brain electrical potential oscillations, we compared each examinee's time series of oscillations in the presence of an MS with the time series of his or her oscillations in the absence of an MS. This approach could be regarded as a personal one. Using this approach, the EEGs of everyone under investigation were measured both in the presence of an MS and in the absence of one.

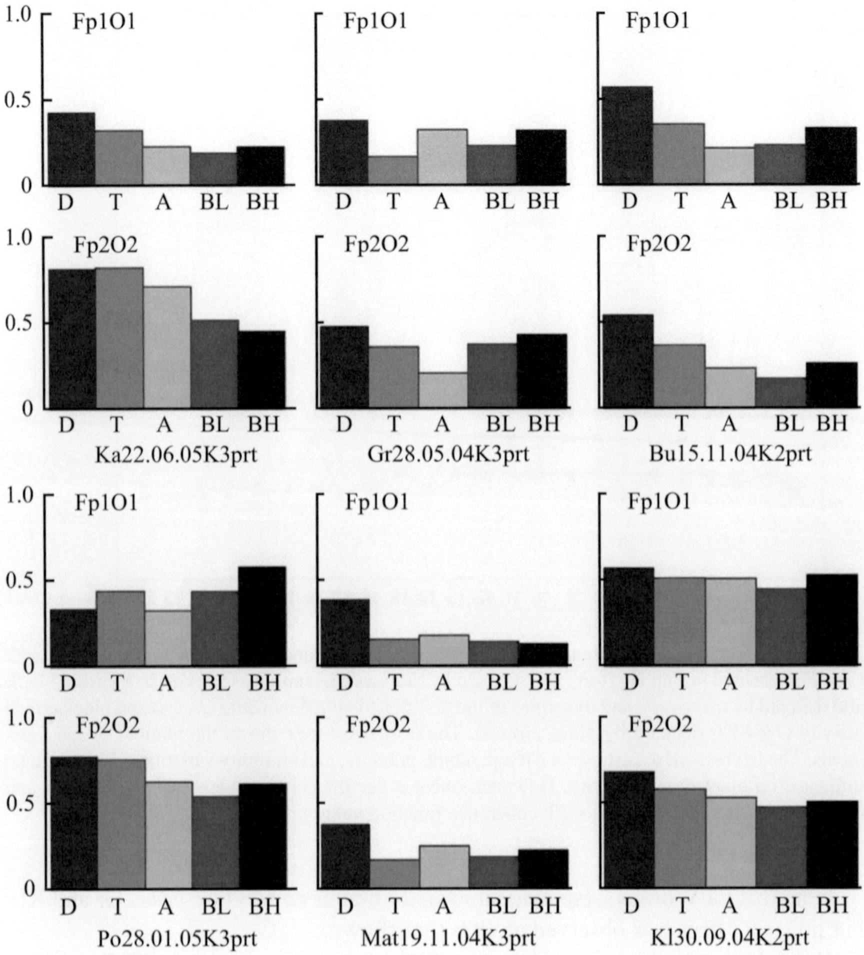

Fig. 6.9 Coherence of brain rhythms (D, T, A, BL, BH) in the frontal polar and occipital regions, determined from electroencephalograms of healthy young volunteers during a computer proofreading test (*prt*) in quiet geomagnetic conditions (geomagnetic index (K_p) = 2 or 3); the *upper diagram* for each examinee (Ka, Gr, . . . Kl) pertains to the left hemisphere (*Fp1O1*) and the *lower diagram* pertains to the right hemisphere (*Fp2O2*)

On the other hand, we had two sets of time series of oscillations: set A of time series measured without the influence of an MS (background recordings), and set B of time series measured in the presence of an MS (storm recordings), regardless of individuals. By this approach, every EEG (of an appropriate quality from the viewpoint of EEG practical standards) could be included in set A or set B and therefore part of the investigation. This approach could be regarded as a sets approach. In other words, a wider total set of EEGs could be used in this case. We could show that there was a number (which we denoted as "FLC"—i.e., the fraction

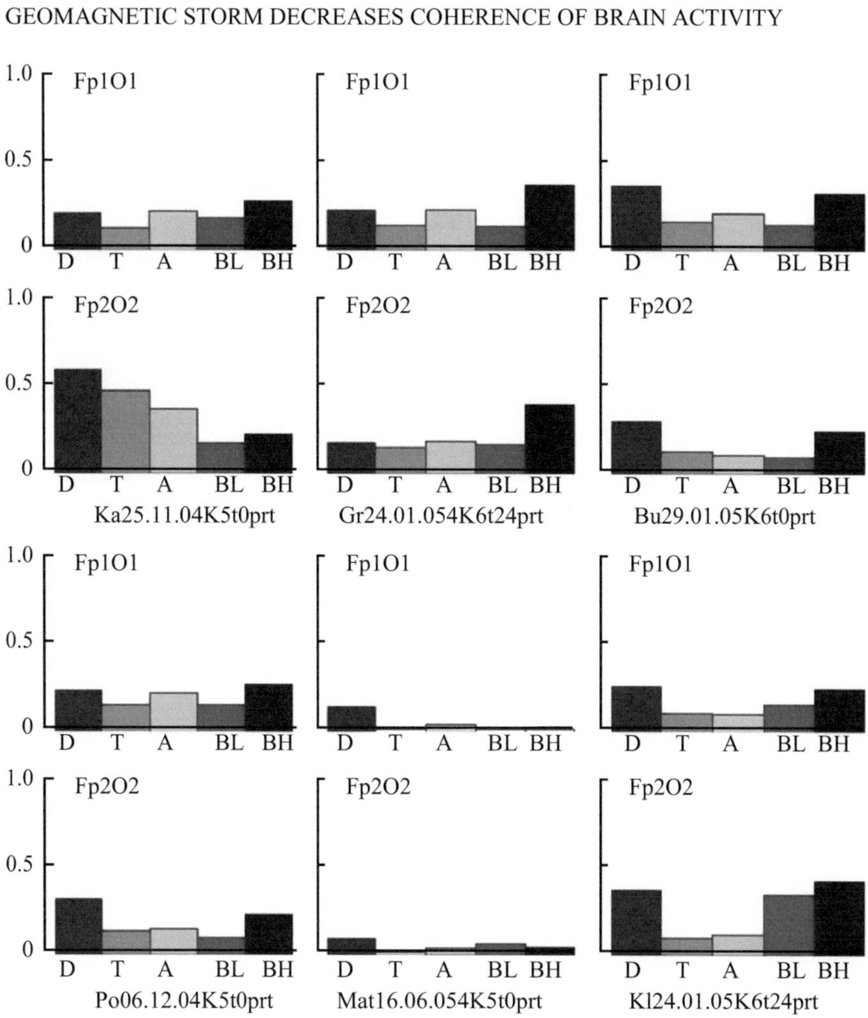

Fig. 6.10 Coherence of brain rhythms (D, T, A, BL, BH) in the frontal polar and occipital regions determined from electroencephalograms of healthy young volunteers during a computer proofreading test (*prt*) during (*t0*) or a day after (*t24*) a magnetic storm (geomagnetic index (K_p) = 5 or 6); the *upper diagram* for each examinee (Ka, Gr, . . . Kl) pertains to the left hemisphere (*Fp1O1*) and the *lower diagram* pertains to the right hemisphere (*Fp2O2*)

of cases of low coherence; see below) that could be determined for both sets A and B and was found to change distinctively under the influence of an MS.

Thus, we needed to distinguish two sets of 24 (set A in our case) and 26 (set B) depersonalized entries. As indicated in Table 6.1, the formal arithmetic means of the CFVs for these sets differed by a factor of two, corroborating the overall effect of the MS revealed in the personal approach.

Table 6.1 Mean coherence (theta rhythm, frontal pole/occipital (Fp/O)) and fraction of cases of low coherence (FLC) at specified coherence function value (CFV) thresholds (Th) in quiet conditions and in magnetic storm conditions

Geomagnetic conditions	Mean CFV	FLC Th 0.3	FLC Th 0.2
Quiet ($K < 5$)	0.4	10/24	5/24
Storm ($K \geq 5$)	0.2	24/26	22/26

Further, we analyzed the composition of the two sets by choosing an arbitrary threshold (Th) and determining the FLC, i.e., the fraction of cases with CF $\leq$ Th out of the total number of entries in the set. These results are shown in Table 6.1. Thus, at a threshold placed simply between the mean CFVs of the sets (Th = 0.3) we obtained an FLC of 10/24 in quiet weather, versus 24/26 in MS conditions. At Th = 0.2 (more rigorous in respect of selectivity), there were only 5/24 low-CFV cases in quiet conditions but as many as 22/26 cases in MS conditions. Here the denominator indicates the whole number of cases considered. The algorithm allowed us to distinguish between the set of EEGs recorded in the absence of an MS and the set recorded under the influence of an MS.

Formally speaking, the numbers Th = 0.3 and Th = 0.2 were the parameters of the algorithm constructed by the authors to find out the qualitative difference between the set of EEG recordings in the absence of an MS and the set of recordings under the influence of an MS. The algorithm was effective in distinguishing the sets of EEG records in the absence and in the presence of an MS. The values of the algorithm's parameters were chosen by us on the basis of the experience of the EEG investigations. Certainly, it was possible to apply all of the values Th = 0.1, 0.2, 0.3, and so on—say, up to 0.9—and thus determine the effective parameter value to distinguish the sets of EEG records in quiet and disturbed geomagnetic conditions.

6.5 Conclusion

In this chapter we have shown that two ways of choosing the threshold parameter Th (Th = 0.3, Th = 0.2) of the algorithm for distinguishing two sets of electroencephalographic records, i.e., one in the absence of a magnetic storm (MS) and the other under the influence of an MS, yielded consistent results. From these results, one can see that the coherence of the frontal polar and occipital theta rhythm brain bioelectrical oscillations in healthy persons at work on a computer was markedly decreased during (or even a day after) a moderate MS, whereas no deviations in the blood pressure, respiratory rate, and or electrocardiographic features were registered. Thereby, the goal formulated in the Introduction of this chapter was attained.

References

Aamodt, S., & Wang, S. (2008). *Welcome to your brain*. London, Bloomsbury.

Babaev, E. S., & Allahverdiyeva, A. A. (2005). *Geomagnetic storms and their influence on the human brain functional state*. Revista CENIC Ciencias Biologicas, 36 (No especial).

Breus, T. K., & Rappoport, S. I. (2003). *Magnetic storms, medical, biological, and geophysical aspects* (In Russian: Magnytnyie buri, medikobiologycheskye I geophysychyskie aspekty). Moscow: Sovyetskyi Sport.

Chyzhevsky, A. L. (1995). *Cosmic pulse of life* (In Russian: Kosmichesky puls zhizny). Moscow: Mysl.

Haken, H. (1996). *Principles of brain functioning. A synergetic approach to brain activity, behavior and cognition*. Heidelberg: Springer.

Merzlyy, A. M., Poskotinova, L. V., Demin, D. B., & Krivonogova, E. V. (2012). *The results of experiment on the influence of earth's magnetic field on the parameters of brain electrical activity*. In International conference on space weather effects on humans: In space and on earth. Book of abstracts, SWH 2.24. Moscow: Space Research Institute of RAS.

Novik, O. B., & Smirnov, F. A. (2007). *A magnetic storm decreases the coherence function of the time series of oscillations of electric potentials of the cerebral cortex during implementation of the proof-reading test*. In Fifth international aerospace congress (IAC-06) proceedings (pp. 361–364). Moscow: Russian Space Agency.

Novik, O. B., & Smirnov, F. A. (2008a). A magnetic storm affects electric potentials of a human brain. *Sun-terrestrial Physics, 12*, 352–353. (Original Russian journal title: Солнечно-земная физика).

Novik, O. B., & Smirnov, F. A. (2008b). *Magnetic storm decreases coherence of oscillations of electric potentials of a human brain*. In International conference GEOCOSMOS – 2008 (pp. 115–116). Book of abstracts, S.-Petersburg.

Novik, O. B., & Smirnov, F. A. (2015). *The brain reaction on geomagnetic storms*. In 11th international scientific and practical conference on manned space flights (pp. 449–450). Oral presentation, Book of abstracts (CD), Yu, A. Gagarin Research & Test Cosmonaut Training Center, Zvezdniy Gorodok, Moscow region. (In Russian: Реакция головного мозга на геомагнитные бури. 11-ая Международная Научно-Практическая Конференция «Пилотируемые полеты в космос», Устный доклад, Тезисы доклада (СД), стр. 449–450. ФГБУ «НИИ ЦПК им. Ю.А. Гагарина», Звездный городок, Московская область).

Pobachenko, S. V., & Kolesnik, A. G. (2012). *Comparative estimation of dynamics of human brain activity in different gelio-geophysics conditions according around the clock monitoring*. In International conference on space weather effects on humans: in space and on earth. Book of Abstracts, SWH 2.24. Moscow: Space Research Institute of RAS.

Smirnov, F. A. (2007). *Magnetic storm is decreasing coherence of electric potentials of the brain when it is functioning*. In International conference on international heliophysical year 2007. New insight into solar-terrestrial physics (IHY2007-NITSP), Book of abstracts. Moscow: Zvenigorod.

Vladymyrsky, B. M., Temuryantz, N. A., & Martynyuk, V. S. (2004). *Cosmic weather and our life*. (In Russian: Kosmicheskaya pogoda i nasha zhizn. Fryazino: Vek 2, 2004. p. 217).

Chapter 7
Elements of Cardiac Anatomy and Physiology

Abstract This chapter is devoted to a description of the anatomy and physiology of the heart at the simplest level possible to facilitate understanding of our new findings in this area—namely the numerical characteristics of heart functioning that are influenced by cosmic weather (see Chap. 4)—for readers unfamiliar with the structure and functions of the heart. In instances where a more detailed analysis of the cardiovascular system is desirable, we recommend that readers study the works included in the reference list of this chapter.

Keywords Atrium · Ventricle · Myocardia · Peacemaker · Depolarization · Repolarization · Automaticity of the heart · Purkinje cells · Myocyte

7.1 General Anatomy and Physiology of the Heart

The heart is a hollow muscular organ, usually located in the left part of the chest. The main function of the heart is pumping of blood throughout the organism to transport gases and nutritional substances to the organs and tissues, and to remove substances produced as a result of metabolism. The heart contains four chambers: two atria and two ventricles. The right atrium and ventricle are used for transport of venous blood. The blood enters the right atrium through the vena cava. The venous blood flows for some distance from the organs, which is why venous blood transports large amounts of carbon dioxide and only small amounts of nutritional substances. The blood enters the right atrium during relaxation of the atria as a result of contraction of the walls of the veins, causing the flow of blood from the organs to the heart. During contraction of the right atrium, blood is pumped from the atrium into the ventricles through the tricuspid valve. During the contraction of the ventricles, the blood passes through the bicuspid valve to the pulmonary arteries. Then the blood goes to the lungs. In the lungs, there is an exchange of gases between the blood and the external medium; as a result, the blood is transformed from its venous form into its arterial form. Then the blood flows along the pulmonary vein to enter the left atrium, from which it passes into the left ventricle. By contraction of the left ventricle, the blood is moved through

O. Novik et al., *Electromagnetic Geophysical Fields*,
https://doi.org/10.1007/978-3-319-98461-2_7

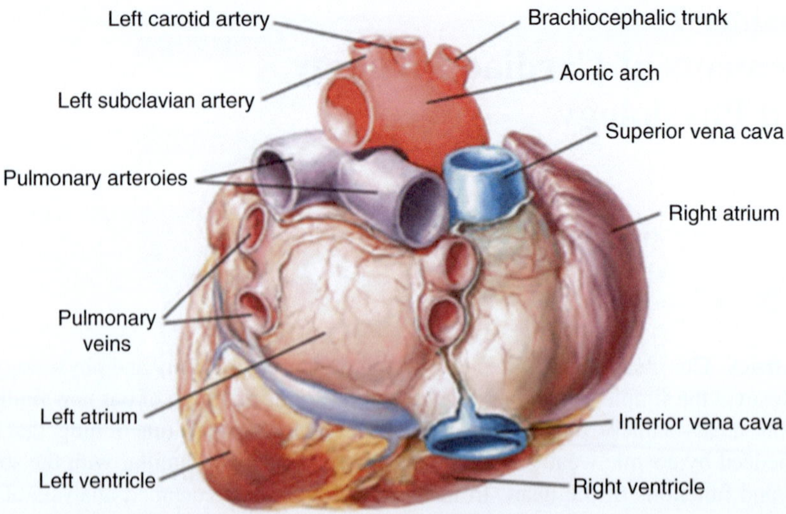

Fig. 7.1 The great vessels, viewed from the posterior side of the heart

the aorta to the organs and tissues. After enrichment of organs and tissues by the oxygen in the arterial blood, the blood enters the veins ones more and flows to the heart. After this final stage, the cycle is repeating again. Figure 7.1 shows the heart and the basic anatomical particularities of its structure. The heart consists of three layers: the pericardium (the connective tissue membrane covering the heart), the myocardium (the muscular layer of the heart, responsible for the motor activity of the heart), and the endocardium (the internal layer of the heart) (Baevsky et al. 2002; Keener and Sneyd 2009; Lippincott 2012; Mohrman and Heller 2014; Zhilin et al. 2014).

The heart has its own system of blood circulation (Fig. 7.2). This system is supplied by the myocardium with oxygen and nutritious substances, which are necessary to realize the contractile function of the myocardium. The coronary arteries and coronary veins are the basic main vessels of the heart. The coronal artery begins at the exit of the aorta from the left ventricle and then it goes to the base of the heart, surrounding it from both sides. The coronary arteries are branched, and the branches go to the top of the heart; as a result, they form the arterial system of the heart. The arteries then branch into smaller vessels, which supply blood to the myocardium, then veins join and form the coronary sinus. The end of this vein is located near the right atrium (Fig. 7.3).

The heart of a human is able to contract independently and spontaneously without participation of the nervous system. This ability is known as the automaticity of the heart. The capacity for automaticity is the result of activity of a small group of cells, located near the venous sinus, near the right atrium. This group of cells is known as the sinoatrial (SA) node. The cells in this group are called pacemakers, or rhythm generators. The frequency of functioning of these cells is equal to the pulse of the

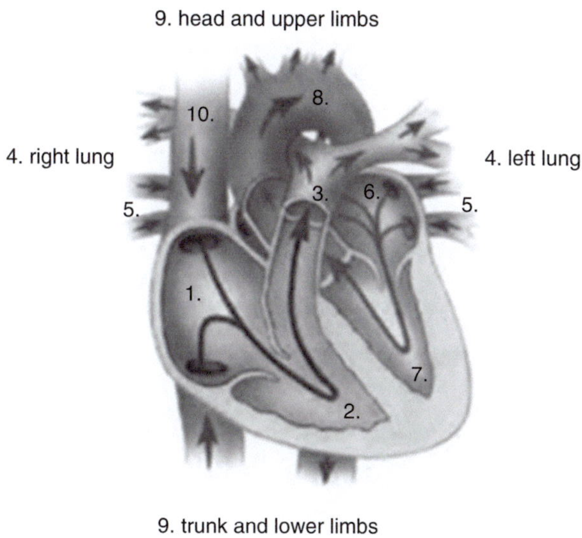

Fig. 7.2 Cardiac circulation. Blood collected in the right atrium is then pumped into the right ventricle. On contraction of the right ventricle, blood passes through the pulmonary trunk and arteries to the lungs. The left atrium pumps the blood into the left ventricle. Contraction of the left ventricle sends the blood through the aortic artery to all tissues in the body. The release of oxygen in exchange for carbon dioxide occurs through capillaries in the tissues. Return of oxygen-poor blood occurs through the superior and inferior vena cava into the right atrium. Note that the unidirectional flow of blood through the heart is accomplished by valves

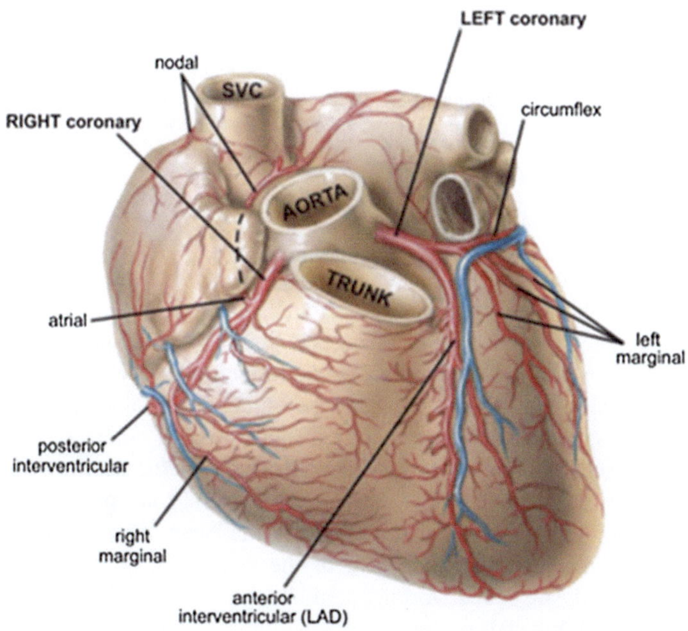

Fig. 7.3 Vascular supply to the heart. Arterial supply to the heart occurs via the right and left coronary arteries and their branches. Venous drainage occurs via the cardiac veins. *LAD* left anterior descending coronary artery, *SVC* superior vena cava

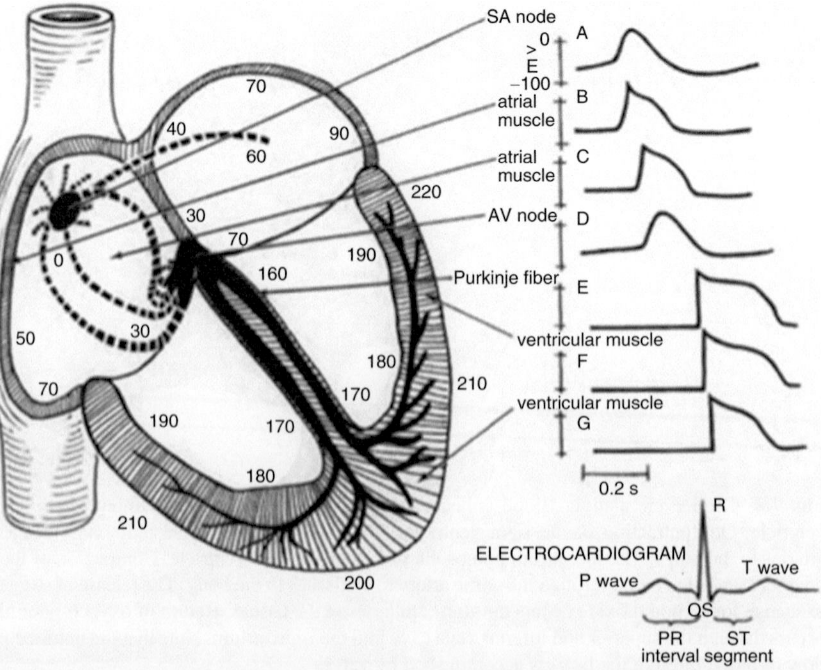

Fig. 7.4 Predominant conduction pathways in the heart and relative times (in ms) when cells in these various regions become activated following an initial depolarization within the sinoatrial (*SA*) node. The SA and atrioventricular (*AV*) nodal cells have similarly shaped action potentials. The nonpacemaker atrial cells elicit action potentials with a shape that is intermediate between those of the slow-response (nodal) cells and fast-response cells (e.g., ventricular myocytes). The ventricular cells elicit fast-response-type action potentials; however, their durations vary in length. Because of the rapid excitation within the Purkinje fiber system, the initiation of depolarization of the ventricular myocytes occurs within 30–40 ms and is recorded as the QRS complex on an electrocardiogram

human during conditions of rest, i.e., from 60 to 70 beats per minute. The ability of these cells to spontaneously generate a rhythm is explained by their inability to hold the transmembrane potential without changes, and this is the reason for the phenomenon of slow spontaneous depolarization of the membrane. When depolarization occurs at the threshold level, the action potential arises, and this potential is propagated through the whole myocardium and makes it contract. It should be mentioned here that this type of cell (pacemaker cell) is a particular form of muscular cell (i.e., an atypical cardiomyocyte). There are similar cells in a particular system of the heart known as the conduction system of the heart. Figure 7.4 shows the locations of the elements and pathways of the conduction system of the heart. Typical muscle cells are called working cardiomyocytes.

Electrochemical transfer of the generated impulse between deferent muscle elements develops with participation of particular types of cell contact (desmosomes) and intercalated disks. The velocity of propagation of the impulse is too

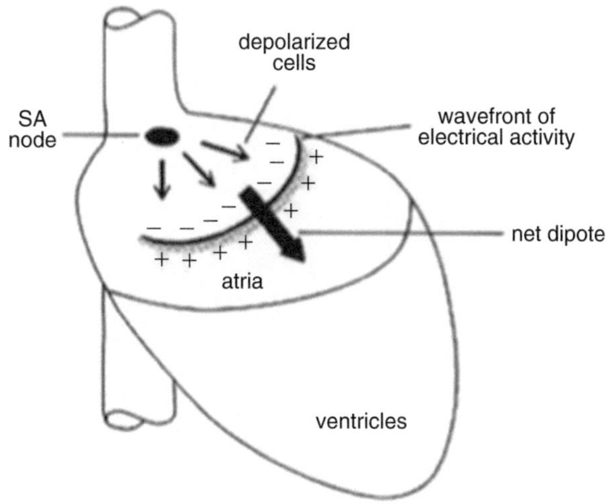

Fig. 7.5 After conduction begins at the sinoatrial (*SA*) node, cells in the atria begin to depolarize. This creates an electrical wave front, which moves down toward the ventricles, with polarized cells at the front, followed by depolarized cells behind. The separation of charge results in a dipole across the heart (the *large black arrow* shows its direction)

slow. This is the reason for the impossibility of fast depolarization of the myocardium as a whole. Time is needed for propagation of the wave of depolarization from the SA node along the His's bundl and Purkinje fibers to the apex of the heart (Fig. 7.5), and then after a change in its direction, for activation of the process of depolarization of the whole myocardium. A delay in the transfer of the impulse takes place, as well, between separate cardiomyocytes in the myocardium. So, the resulting velocity of propagation of the impulse is of the order of 1–4 m/s. The wave of depolarization forms as a result of activity of the SA node, and initially the wave of depolarization is directed from the base of the heart to its apex. So the resulting vector electromagnetic field (EMF) of induction is directed to the apex of the heart. After the arrival of the wave of depolarization at the apex of the heart, the myocardium begins to participate in the process of depolarization and forms a stronger wave of depolarization, and this wave is directed from the apex of the heart to its base. In this case, the vector of EMF induction is directed to the base of the heart. After the end of the process of depolarization of the myocardium, the process of repolarization begins, with the vector of EMF induction directed to the apex of the heart, once more. After the repolarization, the myocardia is going over in its usual state. The trajectory of the motion of the vector of the EMF induction at different moments of the cardiac cycle is called the vector loop. An electrocardiogram (ECG) is a projection of the resulting vector of the EMF of induction to the axis of the ECG leads.

Every wave of the ECG is a fixed-in-time projection of the resulting vector of EMF of induction on the axis of the lead (Fig. 7.6). The P wave is the result of depolarization of the atria, the QRS complex is the result of depolarization of the ventricles, and the T wave arises from repolarization of the myocardium.

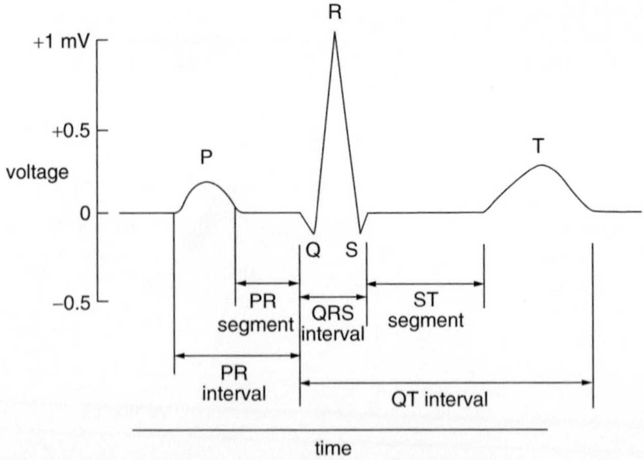

Fig. 7.6 A typical electrocardiographic waveform for one cardiac cycle, measured from the lead II position. The P wave denotes atrial depolarization, the QRS indicates ventricular depolarization, and the T wave denotes ventricular repolarization. The events on the waveform occur on a scale of hundreds of milliseconds

7.2 Glossary

Acetylcholine – Neurotransmitter for nerve signals in the blood, which increases arterial blood pressure, the heart rate, and the heart's contraction force (Picciotto et al. 2002).

Acidosis – Phenomenon arising from delivery of acetic products of metabolism (organic and nonorganic acids, or hydrogen ions) in the blood and extracellular liquid. In extreme cases, acidosis causes changes in the values of basic physiological constants and salt metabolism, potentially with lethal results (Kraut et al. 2010).

Afferent fibers – Nerve fibers transferring nerve signals from peripheral organs or skeletal muscles to the central nervous system (Wikibooks Contributors 2018b).

Alpha adrenoreceptors – Group of adrenoreceptors in organs (cf. beta-adrenergic receptors). Alpha adrenoreceptors are under the influence of adrenaline and noradrenaline. Activation of alpha adrenoreceptors causes decreases in vascular diameter, vascular permeability, and arterial blood pressure, and production of adrenaline and noradrenaline (Mathiazhagan 2013).

Atrioventricular (AV) node – Node located near the base of the partition between the right and left ventricles. It is slightly displaced toward the lower boundary of the atria. The AV node is an additional center of control of cardiac rhythm. The AV node has an impulse frequency of 90–100 pulses per minute, higher than that of the sinoatrial node. The AV node is connected with the sinoatrial node by conduction paths; the letter are path of the conduction system of the heart.

Normally the functioning of the AV node is suppressed by the sinoatrial node, but in the case of a wrong functioning (malfunction) AV node might be the reason of the destruction of the cardiac rhythm (Iaizzo 2005).

Autonomic nervous system (ANS) – Branch of the nervous system that controls the processes of the inner organs. The highest center of the ANS is the thalamus. The ANS controls the frequency of cardiac contractions and intraorgan and interorgan reflexes (Wikibooks Contributors 2018b).

Baroreceptors – Group of receptors for mechanical signals (Martin 2010).

Beta-adrenergic receptors – Receptors located in the walls of the vasculature of the inner organs. These receptors are activated by interaction with adrenaline and noradrenaline. Activation of these receptors increases organ function, arterial blood pressure, the heart rate, and the force of cardiac contractions (Baroffio et al. 2010).

Ca^{2+} ATPase – Ferment that splits molecules of adenosine triphosphate (ATP), producing energy. This ferment, which initializes the contraction of muscles (Antony et al. 2003), is activated by calcium ions released from the sarcoplasmic reticulum through ryanodine channels.

cAMP – Cyclic adenosine monophosphate. This substance is responsible for opening of ion channels and initialization of transmembrane ion currents (Hodavance 2016).

Catecholamine – Group of organic substances regulating the central nervous system and peripheral organs. Catecholamines are neurotransmitters. Like all hormone-like substances, catecholamines take part in local regulation of populations of cells of tissues and their interactions. Typical examples are adrenaline, noradrenaline, and epinephrine (Eisenhofer et al. 2004).

Chemoreceptors – Group of receptors responsible for receiving information about the chemical or electrolytic state of the organism (Moreira et al. 2011).

Cytosol – Liquid part of the cellular medium. Its main function is to solve large organic molecules (polysaccharides, polypeptides, organic acids, etc.). The cytosol contains large molecules of composed proteins; some of them are cellular ferments. Proteins in the cytosol may play a part in the cytoskeleton (Wikibooks Contributors 2018a).

Desmosome – A type of cellular contact joining cells in tissue. There are different kinds of cellular contact, due to the necessity to create different types of interactions between cells. Transfer of interactions is realized by participation of the regulatory chemicals, ion currents, and electrical fields. Desmosomes are created by numerous invaginations and protrusions of plasmatic membranes of neighboring cells, in such a way that a membrane protrusion on one cell comes into contact with a membrane invagination on another cell. As a result, one can see the close mechanical contact. The ion channels of one cell come into contact with the ion channels of another cell, creating an ion transport system, so electrical charges are able to penetrate from one cell to another cell without attenuating, and the tissue becomes a united element from the electrical point of view (Patel 2014).

Efferent fibers – These are also known as motor nerve fibers. Efferent fibers are a part of the nervous system, which conduct impulses from the central nervous system to the organs, including vascular and skeletal muscle, and control motor activity in humans (Wikibooks Contributors 2018b).

G-protein – Part of the system that transfers information from the membrane in the intercellular medium. G-protein is able to initialize the cascade of chemical reactions causing changes in cellular metabolism and increasing the synthesis of new proteins and oxidative processes in a cell. Moreover, G-protein is able to open a number of chemo-dependent ion channels and activate membrane transfer with the help of proteins transporting simple organic chemicals, e.g., glucose and amino acids (Kobilka 2006).

Inotropic effect – Effect reflecting the force of cardiac contraction and its frequency. This effect may be negative (decreasing the frequency and force of cardiac contractions) or positive (increasing the frequency and force of cardiac contractions) (Iaizzo 2005).

Kallikrein–kinin system – System controlling pain activity and decreasing arterial blood pressure and the cardiac contraction rate (Farmer 1997).

Muscarinic receptors – One of the receptor types that transfer nerve signals through the chemical synapsis. These receptors belong to the group of receptors that transfer signals with the help of G-proteins. Muscarinic receptors are activated by muscarine and blocked by atropine (Leach et al. 2012).

Phosphorylation of calcium membrane channels – Chemical reaction resulting in coupling of residual phosphoric acid with a molecule of protein, which is the calcium membrane channel. This process opens the channel and initializes the incoming calcium transmembrane current.

Purkinje cell – A kind of atypical cardiomyocyte located in the Purkinje fibers, which are in electrical contact with His's bundl. They, in turn, are in contact with the atrioventricular node, forming the general conduction system of the heart. Purkinje fibers are the center of cardiac rhythm. In normal conditions, the cardiac rhythm is attenuated by the sinoatrial and atrioventricular nodes. The rhythm of the Purkinje cells may be measured by a hard block of the activity of all generators of cardiac rhythm. Let us note here that the rhythm of the Purkinje cells is propagated only at the top of the heart (Llinás and Sugumori 1992).

Refractory phase – This term denotes the state of a cell of excitable tissue arising from a submaximal or maximal response of the cell to an acting factor. During this phase, the cell demonstrates no reaction, or only a weak reaction, to an external factor. There are absolute and relative refractory phases. In a phase of absolute refractoriness the cell cannot react to an external factor, whereas in a phase of relative refractoriness the cell can react weakly. It is important to differentiate between a refractory phase and the process of the neuron braking (Wikibooks Contributors 2018b).

Ryanodine receptors – A type of potential-dependent calcium channel. The gates of ryanodine receptors are sensitive to changes in transmembrane potential. These channels are responsible for raising the incoming calcium current through changes in the transmembrane potential of the muscle cell. The flow of calcium ions from these channels initiates the process of contraction of a muscle (Petegen 2012).

Sarcoplasmic reticulum – According to cytological terminology, this is a cell organelle—namely an analog of the endoplasmic reticulum. The sarcoplasmic reticulum is needed for synthesis and accumulation of the chemicals produced in a cell. There is the smooth endoplasmic reticulum, and there is the rough endoplasmic reticulum. The term *rough* indicates the presence of a large amount of ribosomes on the membrane of the reticulum. This sort of reticulum synthesizes proteins. The smooth endoplasmic reticulum is responsible, mostly, for synthesis of nonprotein organic substances and their accumulation. The endoplasmic reticulum produces lysosomes and cell inclusions (Martonosi et al. 1967).

Sinoatrial (SA) node – Part of the conduction system of the heart, located between the right atrium and right ventricle, and the aortic arch. It is the main center of control of the heart rhythm, generating impulses with a frequency of 60–70 impulses per minute, and is responsible for the functioning of the heart during rest (Boyett et al. 2000).

Sympathetic nervous system – Branch of the nervous system that causes excitation effects in organs. Excitation of the sympathetic nerve system increases arterial blood pressure, the heart rate, peristalsis of the digestive tract, etc. (Gordan et al. 2015).

Tetanic contraction – A type of muscle contraction. There are two types of tetanic contraction: fused and unfused. Tetanic contraction of muscles occurs as a response to strong stimulation of a muscle by the nervous system or an external factor. The occurrence of the unfused tetanic contraction is explained by a specific basic mechanism of the process of muscle contraction (Pinnell et al. 2007).

Vagal stimulation – This is a physiological term denoting enhancement of the function of an organ or tissue, caused by an impact on the vagus nerve. The vagus nerve includes motor and sensory fibers. The motor fibers can cause inhibition or excitation. The vagus nerve is able to increase the heart rate and ventricle filling, or it can cause the opposite effects. In most cases, vagal influences inhibit cardiac activity, including slowing the heart rate. However, enhancement of cardiac activity is possible through an increase in sympathetic nerve activity (Iaizzo 2005).

Verapamil – This chemical plays a part in blocking calcium ion channels. Verapamil inhibits the calcium current through the membrane of the cardiomyocyte.

7.3 Cardiac Muscle Description

The functional unit of cardiac muscle is a cell called a cardiomyocyte. Each cardiomyocyte is surrounded by a cell membrane called the sarcolemma and contains one nucleus. The cells are packed with mitochondria to provide a steady supply of adenosine triphosphate (ATP), which is required to sustain cardiac contraction. As with skeletal muscle, cardiomyocytes contain the contractile proteins actin (thin filaments) and myosin (thick filaments) together with the regulatory proteins troponin and tropomyosin (Fig. 7.7). Cardiomyocytes are joined end to end at structures known as intercalated disks. The cell membranes form a number of parallel folds and are tightly held together by desmosomes. This results in strong cell-to-cell cohesion, allowing the contraction of one myocyte to be transmitted axially to the next. Gap junctions exist between cardiac muscle cells, providing low resistance pathways for the spread of excitation from one cell to another.

Cardiomyocytes can be divided into work cells and pacemaker cells. The work cells have a large stable resting membrane potential and display a prolonged action potential with a plateau phase (Fig. 7.8). The pacemaker cells have a smaller unstable resting potential and spontaneously depolarize, generating the intrinsic electrical activity of the heart. Pacemaker cells are found in the SA and atrioventricular (AV) nodes. The cells of the His bundle and some Purkinje cells are also capable of spontaneous firing.

The cardiac action potential is very different from that seen in nerves. It has a prolonged plateau phase lasting around 300 ms, compared with 1 ms in nerves. The cardiac action potential has five phases, as shown in Fig. 7.8. During phase 0, membrane permeability to potassium decreases and fast sodium channels open, producing rapid depolarization from 290 mV to 10 mV. During phase 1, there is partial repolarization, because of a decrease in sodium permeability. Phase 2 is the plateau phase of the cardiac action potential. Membrane permeability to calcium increases during this phase, maintaining depolarization and prolonging the action potential. Membrane permeability to calcium decreases somewhat toward the end of phase 2, and the plateau is partially maintained by an inward sodium current. Sodium

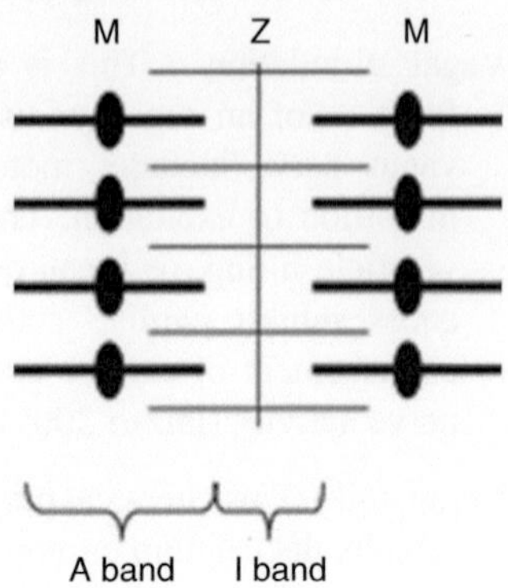

Fig. 7.7 Contractile unit in the cardiac muscle. Actin filaments are anchored at the Z-line and myosin filaments are connected at the M-line. A sarcomere extends from one Z-line to the next

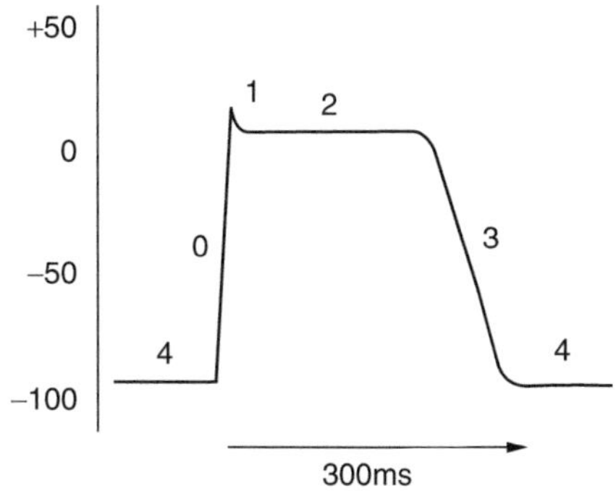

Fig. 7.8 Cardiac action potential. Phase 0: depolarization due to the opening of fast sodium channels. Potassium flux also decreases. Phase 1: partial repolarization due to a rapid decrease in sodium ion passage as fast sodium channels close. Phase 2: plateau phase in which the movement of calcium ions out of the cell maintains depolarization. Phase 3: repolarization; sodium and calcium channels all close and membrane potential returns to baseline. Phase 4: resting membrane potential (290 mV), resulting from activity of the Na^+/K^+ ATPase pump, which creates a negative intracellular potential because of the exchange of three sodium ions for only two potassium ions

flows into the cell through the sodium–calcium exchanger. The exchanger transfers three sodium ions into the cell in exchange for one calcium ion flowing out, thereby producing a net inward positive current.

As calcium channels inactivate toward the end of the plateau phase, an inward potassium current produces repolarization in phase 3. The resting membrane potential in phase 4 is approximately 290 mV. This is produced mainly by the selective permeability of the cell membrane to potassium and the concentration gradient for potassium that exists across the cell membrane and is close to the Nernst equilibrium potential for potassium. Much of the calcium influx in the plateau phase occurs through L-type (long-opening) calcium channels. Increased activation of L-type channels occurs with catecholamine exposure, while they are blocked by calcium channel antagonists such as verapamil. The cardiac action potential lasts about 300 ms. For the vast majority of this time, the cell is absolutely refractory to further stimulation. In other words, a further action potential will not be generated until repolarization is virtually complete. This prevents tetanic contraction from occurring. If a supramaximal stimulus occurs during the relative refractory period, the resultant action potential has a slower rate of depolarization and is of smaller amplitude than normal, producing a much weaker contraction than normal.

The pacemaker potential is seen in cells in the SA and AV nodes. As shown in Fig. 7.9, it differs from the action potential of cardiomyocytes in that phases 1 and 2 are absent. The heart displays autorhythmicity: a denervated heart (such as a transplanted heart) continues to contract spontaneously. Pacemaker cells do not have a stable resting action potential, and it is the spontaneous depolarization of the pacemaker potential that gives the heart its autorhythmicity. The pacemaker potential is produced by a decrease in membrane permeability to potassium, a slow

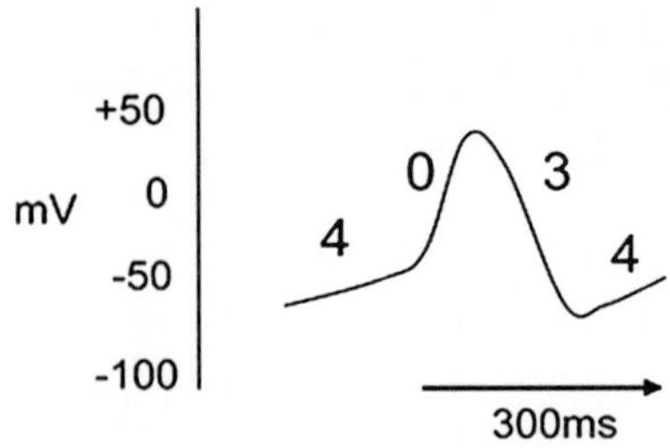

Fig. 7.9 Pacemaker potential. Phases 1 and 2 do not occur. Phase 4: pacemaker potential. Phase 0: depolarization due to voltage-gated channel opening (as opposed to ventricular muscle action potential). Phase 3: repolarization due to a decrease in potassium

inward current due to calcium influx via T-type (transient) calcium channels, and an increased sodium current due to sodium–calcium exchange. Once the threshold potential is reached, L-type calcium channels open, calcium ions enter the cell, and depolarization occurs. In contrast to the cardiomyocyte action potential, there is no inward movement of sodium ions during depolarization. Repolarization (phase 3 of the action potential) occurs because of an increase in potassium permeability. At the SA node, potassium permeability can be further enhanced by vagal stimulation. This has the effect of hyperpolarizing the cell and reducing the rate of firing. Sympathetic stimulation has the opposite effect.

The SA node, other atrial centers, the AV node, and the His bundle all display inherent pacemaker activity. The SA node has the highest rate of spontaneous depolarization and therefore suppresses the other pacemakers. In a denervated heart, the SA node discharges at a rate of approximately 100 times/min. Vagal tone leads to a lower heart rate in healthy subjects at rest. From the SA node, impulses spread throughout the atria to the AV node at a rate of 1 m/s. The AV node is the only means of electrical connection between the atria and the ventricles. Conduction here is slow (approximately 0.05 m/s). This means that the AV node will transmit a maximum of only 220 impulses/min, thereby protecting the ventricles from high rates of atrial depolarization. Depolarization spreads from the AV node to the His bundle in the interventricular septum. The bundle splits into right and left bundle branches, supplying the respective ventricles. The left bundle itself divides into anterior and posterior divisions. From the bundle branches, impulses travel through the ventricular muscle, via a network of Purkinje fibers, at a velocity of 1–4 m/s. The conducting system is arranged so that the apices of the ventricles contract before the bases, propelling blood out of the chambers.

Excitation contraction coupling is the process linking electrical excitation to contraction. Calcium has an essential role in this process; a raised intracellular calcium concentration is the trigger that activates contraction. An understanding of calcium handling is essential to understanding the function of the heart. The intracellular calcium ion concentration in the cardiomyocyte at rest is 0.0001 mM/L and that in the extracellular fluid is 1.2 mM/L. During the plateau phase of the action potential, calcium ions flow down this steep concentration gradient and enter the myocyte. Most of this calcium enters through L-type channels, located primarily at sarcolemmal/sarcoplasmic reticulum junctions. The influx of calcium triggers the release of further calcium from the sarcoplasmic reticulum via ryanodine receptors. This calcium-triggered calcium release differs from that in skeletal muscle, where the action potential triggers calcium release directly. Free intracellular calcium interacts with the C subunit of troponin. This leads to a configuration change in the troponin/tropomyosin complex, allowing actin to interact with myosin. Cross-bridge cycling occurs, leading to a shortening of the sarcomere and resultant muscular contraction. As intracellular calcium concentrations decrease during repolarization, calcium dissociates from troponin as the intracellular calcium concentration decreases, resulting in relaxation. Diastolic relaxation is an active (ATP-dependent) process. Calcium transport out of the cytosol occurs via sarcoplasmic reticulum Ca^{2+} ATPase, through sarcolemmal Na^{+}/Ca^{2+} exchange, via sarcolemmal Ca^{2+} ATPase, and finally by utilizing a mitochondrial Ca^{2+} uniport (index p and + are the equivalent notatoins here). The strength of a contraction may be varied by increasing the amount of free intracellular calcium, by altering the sensitivity of the myofilaments to calcium, or both. The latter occurs during stretching of the myofilaments and is responsible for the Frank–Starling mechanism. Myofilament calcium sensitivity is reduced by acidosis. High concentrations of phosphate and magnesium also impair cardiac function. Catecholamines activate beta-adrenergic receptors in the heart to produce a G-protein-mediated increase in cyclic adenosine monophosphate (cAMP) and enhanced activity of cAMP-dependent protein kinase. This leads to phosphorylation of calcium membrane channels, enhancing calcium entry into the cell. Phosphorylation of myosin also occurs, increasing the rate of cross-bridge cycling. Catecholamines also increase the rate of reuptake of calcium into the sarcoplasmic reticulum, thus aiding relaxation.

7.4 Cardiac Conduction System

Automaticity is the ability of cells to spontaneously depolarize and raise the resting membrane potential past the threshold potential for triggering an action potential. Automaticity normally is a property of cells localized within the SA and AV nodes, as well as the Purkinje fibers. Electrical conduction through the myocardium may be considered at the levels of signal transit along a single myocyte, cell-to-cell conduction, and conduction through the whole organ. The electrical signal passes along a myocyte by incrementally depolarizing the membrane. At rest, an electrical gradient is maintained across the sarcolemmal membrane. When the local membrane potential is depolarized, sodium and calcium channels open, establishing a small region of positive charge on the inner surface of the membrane and leaving a small zone of negative charge extracellularly. This depolarizes the membrane in the immediately adjacent regions and leads to initiation of the depolarizing current in the bordering regions. This forms a self-perpetuating reaction with the spread of depolarization.

The more rapidly the local region of the membrane is able to change its potential, the more rapid the conduction down the length of the myocyte is. Therefore, cells that depend on the rapid sodium current for the upstroke of phase 0 of the action potential conduct the signals rapidly, while cells that are dependent on the slower calcium current for phase 0 conduct signals more slowly. This concept is demonstrated by comparing the conduction velocity in the AV node (calcium-dependent action potentials) with that in atrial and ventricular tissue (sodium channel-dependent). The myocardium is not a perfect syncytium, and conduction is discontinuous. Gap junctions that form a low resistance passageway to allow for the rapid spread of excitation from cell to cell attach the myocytes to one another. It has also been appreciated that the tissue passes electrical current more rapidly along the long axis of the myocytes than transversely. The difference in the conduction velocities axially versus transversely is referred to as anisotropy. The degree of anisotropic conduction varies throughout the myocardium. In the ventricle, the ratio of conduction velocity that is parallel, as opposed to transverse, to the long axis of the cell is approximately 3:1, while along the crista terminal in the atrium, the same ratio is 10:1. On a macroscopic level, once the electrical signal escapes from the SA node, it rapidly spreads throughout the atrium. A debate remains as to whether or not there are anatomically distinct internodal tracts that connect the SA and AV nodes. Histological studies have failed to demonstrate the presence of these tracts. The SA node is located high in the right atrium. In this location it is susceptible to injury at the time of cannulation for cardiopulmonary bypass. The electrical signal activates the right atrium initially with conduction spread into the left atrium preferentially via Bachmann's bundle and along the coronary sinus. The atria and ventricles are electrically isolated from one another by the fibrous AV ring, with electrical continuity provided by the AV node. The compact AV node is located in the right atrium at the apex of the triangle of Koch. The AV node is composed of three regions: the atrionodal (AN), nodal (N), and nodal–His (NH). The cells in these three regions differ in terms of the shapes of their respective action potentials and their conduction velocities. The cells in the AN region have action potentials that are intermediate between those of atrial and SA nodal cells, with a more depolarized maximal diastolic potential, slower phase 0 depolarization, and the presence of phase 4 depolarization. The N cells are similar to the SA nodal cells. The NH cells transition between the N cells and the His bundle with action potentials that reflect this transition. Conduction through the entirety of the AV node is slower than that through the atria or ventricles and is not detectable on a routine surface ECG. The AV node is richly innervated by the autonomic nervous system (ANS) and receives its blood supply from the posterior descending coronary artery. Atrioventricular block may result from interruption of the blood supply, which may result from atherosclerotic disease, as well as from vasospasm following the delivery of radiofrequency energy in the posteroseptal region on the tricuspid valve annulus. After passing through the AV node, the signal passes through the His bundle and enters the Purkinje fibers that divide into the bundle branches. The Purkinje cells have large diameters and preferential end-to-end connections rather than side-to-side connections, both of which lead to accelerated conduction velocities (200 cm/s compared with 5 cm/s in the SA node). The right bundle branch is cord-like and passes to the apex of the right ventricle prior to ramifying into the ventricular mass initially along the moderator band. The left bundle branch is fan-like and initially activates the ventricular septum from the left ventricular side, progressing toward the right. This accounts for the Q waves inscribed in the left lateral precordial leads in D-looped hearts, and the Q waves in the right precordium in L-looped hearts.

7.5 Heart Innervation

The medulla, located in the brain stem above the spinal cord, is a major site in the brain for regulating autonomic nerve outflow to the heart and blood vessels, and is particularly important for short-term feedback regulation of arterial pressure. The medulla contains cell bodies for the two main divisions of the ANS—sympathetic and parasympathetic. The sympathetic nerves exit the medulla and travel down the spinal cord, where they synapse with relatively short preganglionic fibers that travel to, and synapse within, sympathetic ganglia. Postganglionic efferent fibers from the ganglia travel to the heart and vasculature, where they synapse at their target sites. The parasympathetic nerves (vagus nerve, cranial nerve) exit the medulla as long preganglionic efferent fibers that form synapses with short postganglionic fibers within the heart or vascular tissue. The activity of the medullary neurons is modulated by input from peripheral sensors and from other brain regions (see Figs. 7.10 and 7.11).

The nucleus tractus solitarius (NTS) of the medulla receives sensory input from different systemic and central receptors (e.g., baroreceptors and chemoreceptors).

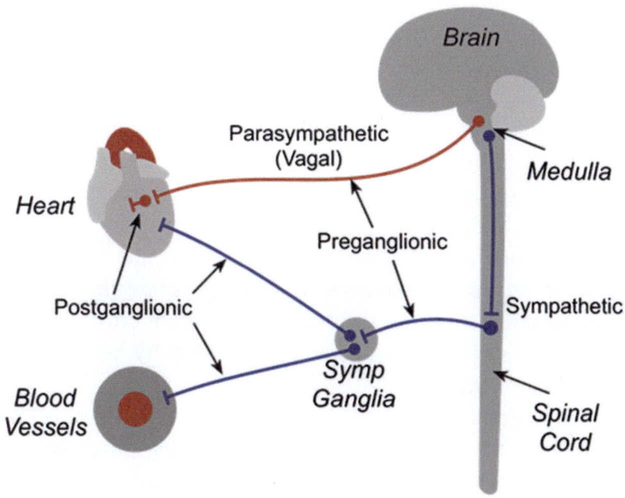

Fig. 7.10 Autonomic innervation of the heart and vasculature

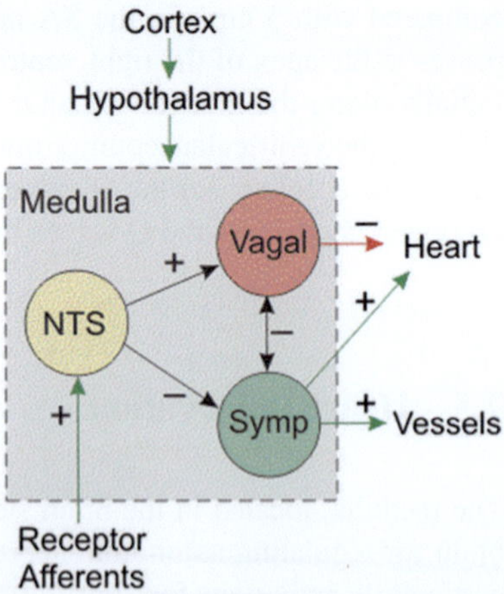

Fig. 7.11 Cardiovascular center: heart control mechanism

Neural connections from the NTS modulate the activity of sympathetic neurons located in the medulla, and the activity of parasympathetic neurons located in the vagal nucleus and nucleus ambiguous, from which arise the parasympathetic vagus nerve (tenth cranial nerve). Neuronal activity in the NTS reciprocally activates vagal neurons and inhibits sympathetic neurons. The medulla also receives information from other brain regions (e.g., the hypothalamus). The hypothalamus and higher centers modify the activity of the medullary centers and are particularly important in stimulating cardiovascular responses to emotion and stress (e.g., exercise, thermal stress). Autonomic outflow from the medulla is divided into sympathetic and parasympathetic (vagal) branches. Efferent fibers of these autonomic nerves travel to the heart and blood vessels, where they modulate the activity of these target organs (Fig. 7.11).

The heart is innervated by vagal and sympathetic fibers. The right vagus nerve primarily innervates the SA node, whereas the left vagus innervates the AV node; however, there can be a significant overlap in the anatomical distribution. Atrial muscle is also innervated by vagal efferents, whereas the ventricular myocardium is only sparsely innervated by vagal efferents. Sympathetic efferent nerves are present throughout the atria (especially in the SA node) and ventricles, including the conduction system of the heart.

Sympathetic stimulation of the heart increases the heart rate (positive chronotropy), inotropy, and conduction velocity (positive dromotropy), whereas parasympathetic stimulation of the heart has the opposite effects. Sympathetic and parasympathetic effects on heart function are mediated by beta-adrenoceptors and muscarinic receptors, respectively.

Sympathetic adrenergic nerves travel along arteries and nerves and are found in the adventitia (outer walls of blood vessels). Varicosities, which are small enlargements along the nerve fibers, are the site of neurotransmitter (norepinephrine) release. Capillaries receive no innervation. Activation of vascular sympathetic nerves causes contraction of the vascular smooth muscle and vasoconstriction of arteries and veins, mediated by alpha-adrenoceptors.

Parasympathetic fibers are associated with blood vessels in certain organs such as the salivary glands and gastrointestinal glands, and in genital erectile tissue. The release of acetylcholine (ACh) from these parasympathetic nerves, which binds to muscarinic ACh receptors, has a direct vasodilatory action (coupled to nitric oxide formation and guanylyl cyclase activation). ACh release can stimulate the release of kallikrein from glandular tissue, which acts upon kininogen to form kinins (e.g., bradykinin). Kinins cause increased capillary permeability and venous constriction, along with arterial vasodilation in specific organs.

7.6 The Electrocardiogram

An ECG is a measure of how the electrical activity of the heart changes over time as action potentials propagate throughout the heart during each cardiac cycle. However, this is not a direct measure of the cellular depolarization and repolarization of the heart; rather, it is a measure of the relative, cumulative magnitude of populations of cells eliciting changes in their membrane potentials at a given point in time; it shows electrical differences across the heart when depolarization and repolarization of these atrial and ventricular cells occur.

For the purposes of an ECG, the human body can be considered a large-volume conductor. It is basically filled with tissues surrounded by a conductive ionic fluid. One can imagine that the heart is suspended inside that conductive medium. During the cardiac cycle, the heart contracts in response to action potentials moving along the chambers of the heart. As they move, there will be one part of the cardiac tissue that is depolarized and another part that is at rest or polarized. This results in a charge separation, or dipole. The dipole causes current flow in the surrounding body fluids between the ends of the heart, resulting in a fluctuating electrical field throughout the body. This is much like the electrical field that would result, for example, if a common battery were suspended in a saltwater solution (an electrically conductive medium). The opposite poles of the battery would cause current flow in the surrounding fluid, creating an electrical field that could be detected by electrodes placed in the solution. A similar electrical field around the heart can be detected using electrodes attached to the skin. The intensity of the voltage that is detected depends on the orientation of the electrodes with respect to that of the dipole ends. The amplitude of the signal is proportional to the mass of tissue involved in creating that dipole at any given time. Use of electrodes on the surface of the skin to detect the voltage of this electrical field is what provides the ECG.

It is important to note, as might be expected, that because the ECG is measured on the skin, any potential differences within the body can have an effect on the electrical field that is detected. This is why it is considered important for diagnostic purposes that, during recording of an ECG, the patient should remain as still as possible. Movements require the use of skeletal muscles, which then contribute to the changes

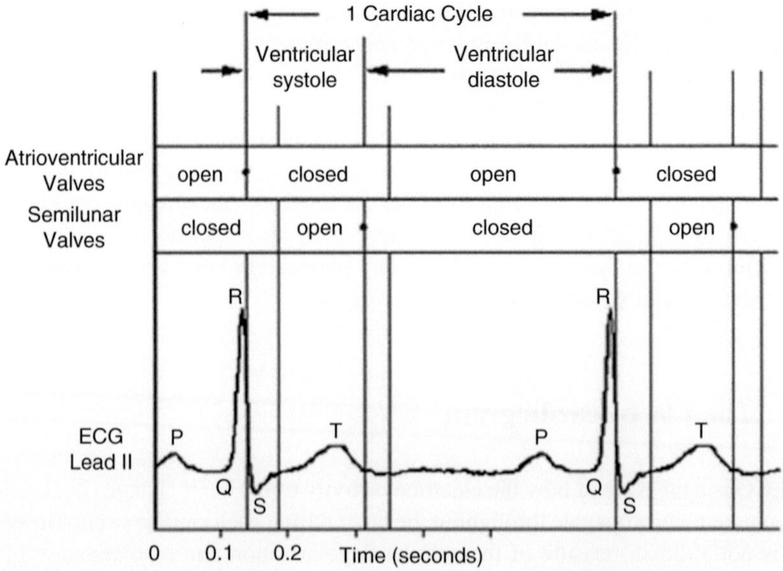

Fig. 7.12 Typical lead II electrocardiogram waveform showing the timing of atrioventricular and semilunar valve activity; during these segments of the cardiac cycle, the ventricles are in systole/diastole

in voltages detected using electrodes on the surface of the body. When the monitored patient is essentially motionless, it is considered a resting ECG.

When an ECG is recorded, a reading of voltage versus time is produced, which is normally displayed as millivolts (mV) and seconds. A typical lead II ECG waveform is shown in Fig. 7.12. For this recording, the negative electrode was placed on the right wrist and the positive electrode was placed on the left ankle, giving the standard lead II ECG. It shows a series of peaks and waves, which corresponds to ventricular or atrial depolarization and repolarization, with each segment of the signal representing a different event associated with the cardiac cycle.

The cardiac cycle begins with the firing of the SA node in the right atrium. This firing is not detected by the surface ECG because the SA node is not composed of an adequately large number of cells to create an electrical potential with a high enough amplitude to be recorded with distal electrodes (signal amplitude is lost as it dissipates through the conductive medium). The atria then depolarize, giving rise to the P wave. This represents the coordinated depolarization of the right and left atria and the onset of atrial contraction. The P wave is normally around 80–100 ms in duration. As the P wave ends, the atria are completely depolarized and are beginning to contract.

The signal then returns to baseline, and action potentials (not large enough to be detected) spread to the AV node and His bundle. Then, roughly 160 ms after the beginning of the P wave, the right and left ventricles begin to depolarize, resulting in what is called the QRS complex, representing the beginning of ventricular contraction, which is around 80 (60–100) ms in duration. Typically, the first negative deflection is the Q wave, the large positive deflection is the R wave, and if there is a negative deflection after the R wave, it is called the S wave. The exact shape of the QRS complex depends on the placement of electrodes from which the signals are recorded.

Simultaneously with the QRS complex, atrial contraction has ended, and the atria are repolarizing. However, the effect of this global atrial repolarization is sufficiently masked by the much larger amount of tissue involved in ventricular depolarization and is thus is not normally detected on the ECG. During ventricular contraction, the ECG signal returns to baseline. The ventricles then repolarize after contraction, giving rise to the T wave. Note that the T wave is normally the last potential detected in the cardiac cycle; thus, it is followed by the P wave of the next cycle, repeating the process.

Of clinical importance in the ECG waveform are several notable parameters (regions), which include the P–R interval, the S–T segment, and the Q–T interval. The P–R interval is measured from the beginning of the P wave to the beginning of the QRS complex and is normally 120–200 ms long. This is basically a measure of the time it takes for an impulse to travel from atrial excitation through the atria, AV node, and remaining fibers of the conduction system. The S–T segment is the period of time when the ventricles are completely depolarized and contracting, and it is measured from the S wave to the beginning of the T wave. The Q–T interval is measured from the beginning of the QRS complex to the end of the T wave; this is the time segment from when the ventricles begin their depolarization to the time when they have repolarized to their resting potentials, and it is normally about 400 ms in duration.

An obvious observation concerning the QRS complex is that it has a much higher peak and shorter duration than either the P or T waves. This is because ventricular depolarization occurs over a greater mass of cardiac tissue (i.e., a greater number of myocytes are depolarizing at the same time); furthermore, the ventricular depolarization is much more synchronized than either atrial depolarization or ventricular repolarization.

It is also very important to note that deflections in the ECG waveform represent changes in electrical activity caused by atrial or ventricular depolarization and repolarization.

One last thing that should be noted relative to the ECG waveform is the occasionally detected potential referred to as the U wave. Its presence is not fully understood, but it is considered by some to be caused by late repolarization of the Purkinje system. If detected, the U wave will appear toward the end of the T wave and have the same polarity (positive deflection). However, it has a much shorter amplitude and usually ascends more rapidly than it descends (unlike the T wave).

Typically, the ECG is measured from the surface of the skin, which can be done by placing two electrodes directly on the skin and reading the potential difference between them. This is possible because these signals are transmitted throughout the body. As stated above, the detected waveform features depend not only on the amount of cardiac tissue involved but also on the orientation of the electrodes with

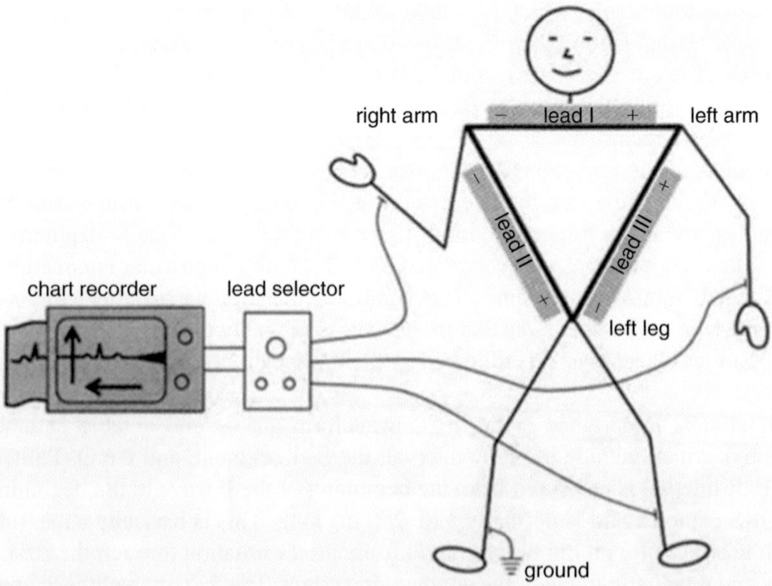

Fig. 7.13 Limb leads are attached to the corners of Einthoven's triangle on the body. Each lead uses two of these locations for a positive and a negative lead. The *plus and minus signs* indicate the orientation of the polarity conventions

respect to the dipole in the heart. In other words, the ECG waveform will look slightly different when measured from different electrode positions, and typically an ECG is obtained using a number of different electrode locations (e.g., limb leads or precordial leads) or configurations (unipolar, bipolar, modified bipolar) which, fortunately, have been standardized by universal application of certain conventions.

The three most commonly employed lead positions used today are referred to as leads I, II, and III. For the purposes of explaining the position of these leads, imagine the torso of the body as an equilateral triangle, as illustrated in Fig. 7.13. This forms what is known as Einthoven's triangle (named after the Dutch scientist who first described it). Electrodes are placed at each of the vertices of the triangle, and a single ECG trace (lead I, II, or III) is measured along the corresponding side of the triangle, using the electrodes at each end. (Because each lead uses one electrode on either side of the heart, leads I, II, and III are also referred to as the bipolar leads.) The plus and minus signs shown in Fig. 7.14 indicate the polarity of each lead measurement and are considered a universal convention.

The limb leads are attached to the corners of Einthoven's triangle on the body. Each lead uses two of these locations for a positive and a negative lead. The plus and minus signs indicate the orientation of the polarity conventions.

As an example, if the lead II ECG trace shows an upward deflection, it would mean that the voltage measured at the left leg (or at the bottom apex of the triangle) is

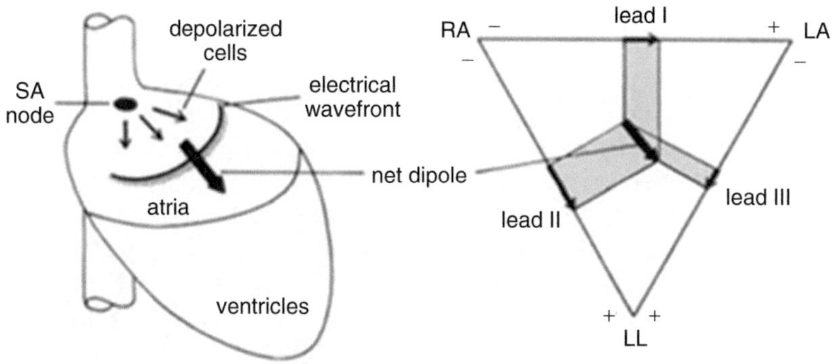

Fig. 7.14 The net dipole occurring in the heart at any one point in time is detected by each lead (I, II, and III) in a different way because of the different orientations of each lead set relative to the dipole in the heart. In this example, the projection of the dipole on all three leads is positive (the *arrow* is pointing toward the positive end of the lead), which gives a positive deflection on the electrocardiogram during the P wave. Furthermore, lead II detects a larger-amplitude signal than lead III does from the same net dipole (i.e., the net dipole projects a *larger arrow* on the lead II side of the triangle than on the lead III side). *LA* left arm, *LL* left leg, *RA* right arm, *SA* sinoatrial

more positive than the voltage measured at the right arm (or at the upper right apex of the triangle). One notable time point at which this happens is during the P wave. Imagine the orientation of the heart as shown in Fig. 7.14, with the action potential propagation across the atria creating a dipole pointed downward and to the left side of the body. This can be represented as an arrow in Fig. 7.14, showing the magnitude and direction of the dipole in the heart. This dipole would, overall, create a more positive voltage reading at the left ankle electrode than at the right wrist electrode, thus eliciting the positive deflection of the P wave on the lead II ECG.

Now, imagine how that same action potential propagation would appear on the other lead placements, leads I and III, if placed at the center of Einthoven's triangle (Fig. 7.14, right side). Each of these lead placements can be thought of as viewing the electrical dipole from three different directions: lead I from the top, lead II from the lower right side of the body, and lead III from the lower left side, all looking at the heart in the frontal plane. In this example, the atrial depolarization creates a dipole that gives a positive deflection for all three leads because the arrow's projection onto each lead (in other words, measuring the cardiac dipole from each lead) results in the positive end of the dipole being pointed more toward the positive end of the lead than the negative end. This is why atrial depolarization (the P wave) appears as a positive deflection for each lead (although the wave magnitude is different in each). Ventricular depolarization, however, is a bit more complicated; briefly, it results in various directions of the Q- and S-wave potentials, depending on which lead trace is utilized for recording.

The direction and magnitude of the overall dipole of the heart at any instant (represented by the arrow in Fig. 7.14, for example) is also known as the heart's electrical axis, which is a vector originating in the center of Einthoven's triangle such that the direction of the dipole is typically assessed in degrees. The convention for this is to use a horizontal line across the top of Einthoven's triangle as 0° and move clockwise downward (pivoting on the negative end of lead I) as the positive direction. It should be noted that the electrical axis is actually changing direction throughout the cardiac cycle as different parts of the heart depolarize/repolarize in different directions. Figure 7.14 shows the dipole spreading across the heart during a typical cardiac cycle, beginning with atrial depolarization. Keep in mind that at certain points, the electrical axis of the heart may give opposite deflections on the various ECG leads. As can be observed in Fig. 7.14, depolarization begins at the SA node in the right atrium, forming the P wave. The atria depolarize downward and to the left, toward the ventricles, followed by a slight delay at the AV node before the ventricles depolarize. The initial depolarization in the ventricles normally occurs on the left side of the septum, creating a dipole pointed slightly down and to the right. This gives a negative deflection of the Q wave for leads I and II; however, it is positive in lead III. Depolarization then continues to spread down the ventricles toward the apex, which is when the most tissue mass is depolarizing at the same time, with the same orientation. This gives the large positive deflection of the R wave for all three leads. Ventricular depolarization then continues to spread through the cardiac wall and finally finishes in the left ventricular lateral wall. This results in a positive deflection for leads I and II; however, lead III shows a lower R-wave amplitude along with a negative S-wave deflection.

After a sustained depolarized period (the S–T segment), the ventricles then repolarize. This occurs anatomically in the opposite direction to depolarization. However, one must keep in mind that the arrow in Fig. 7.14 represents the electrical axis of the heart (or the dipole) and does not necessarily show the direction in which the repolarization wave is moving. Thus, even though the wave is moving from the epicardium to the endocardium (the direction of repolarization), the dipole (and therefore the electrical axis) remains in the same orientation as during depolarization. This explains why the T wave is also a positive deflection on leads I and II and is negative (or nonexistent) on lead III. The ventricles are then repolarized, returning the signal to its baseline potential (value).

During the cardiac cycle, the electrical axis of the heart (viewed from the plane of the limb leads) is always changing in both magnitude and direction. The average of all instantaneous electrical axis vectors gives rise to the mean electrical axis of the heart. Most commonly, it is taken as the average dipole (or electrical axis) direction during the QRS complex, since this is the highest and most synchronized signal on the waveform. Typically, to find the mean electrical axis, area calculations from under the QRS complexes from at least two leads are needed. However, it is more easily and more commonly determined by an estimate using the deflection (positive or negative) and height of the R wave.

References

Antony, N., et al. (2003). The structure of the Ca^{2+}-ATPase of sarcoplasmic reticulum. *Acta Biochimica Polonica, 50*(2), 337–365.

Baevsky, R. M. et al. (2002). *HRV analysis under the usage of different electrocardiography systems (methodical recommendations)*. Committee of Clinic Diagnostic Apparatus and the Committee of New Medical Techniques of Ministry of Health of Russia (protocol № 4 from 11th of April, 2002).

Baroffio, M., et al. (2010). Beta-adrenergic agonists. *Pharmaceuticals, 3*, 1016–1044. https://doi.org/10.3390/ph3041016. 10.01.2019

Boyett, M. R., et al. (2000). Review: The sinoatrial node, a heterogeneous pacemaker structure. *Cardiovascular Research, 47*, 658–687.

Eisenhofer, G., et al. (2004). Catecholamine metabolism: A contemporary view with implications for physiology and medicine. *Pharmacological Review, 56*, 331–349.

Farmer, S. (1997). *The kinin system*. Cambridge: Academic Press.

Gordan, R., et al. (2015). Autonomic and endocrine control of cardiovascular function. *World Journal of Cardiology, 7*(4), 204. https://doi.org/10.4330/wjc.v7.i4.204.

Hodavance, S. Y. (2016). G protein-coupled receptor biased agonism. *Journal of Cardiovascular Pharmacology, 3*, 193–202.

Iaizzo, P. A. (Ed.). (2005). *Handbook of cardiac anatomy, physiology and devices*. Totowa: Humana Press.

Keener, J., & Sneyd, J. (2009). *Mathematical physiology* (2nd ed.). New York: Springer. https://doi.org/10.1007/978-0-387-79388-7.

Kobilka, B. K. (2006). Review of G-protein coupled receptor structure and activation. *Biochemica et Biophysica Acta, 2006*, 794–807.

Kraut, J. A., et al. (2010). Review metabolic acidosis: Pathophysiology, diagnosis and management. *Nature Reviews Nephrology, 6*, 274–285. https://doi.org/10.1038/nrneph.2010.33.

Leach, K., et al. (2012). *Handbook of experimental pharmacology*. Berlin: Springer. https://doi.org/10.1007/978-3-64223274-9_2.

Lippincott, W. (2012). *Cardiovascular physiology concepts* (2nd ed.). Philadelphia: Lippincott Williams & Wilkins.

Llinás, R. R., & Sugumori, M. (1992). The electrophysiology of the cerebellar Purkinje cell revisited. In R. R. Llinás & C. Sotelo (Eds.), *The cerebellum revisited*. New York: Springer. https://doi.org/10.1007/978-1-4612-2840-0_8.

Martin, J .L. (2010). *BDNF in developing aortic baroreceptors and its role in the dendritic morphogenesis of neurons from the nucleus tractus solitarius*. OHSU Digital Commons Scholar Archive. https://digitalcommons.ohsu.edu/cgi/viewcontent.cgi?referer=https://www.google.ru/&httpsredir=1&article=1656&context=etd

Martonosi, A., et al. (1967). The role of phospholipids in the adenosine triphosphatase activity and Ca^{++} transport. *Journal of Biological Chemistry, 243*(1), 61–70.

Mathiazhagan, S. (2013). Alfa adrenergetic receptors: A brief perspective. *IOSR Journal of Dental and Medical Sciences, 7*(4), 16–19.

Mohrman, D. E., & Heller, L. J. (Eds.). (2014). *Cardiovascular physiology* (8th ed.). New York: Lange.

Moreira, T. S., et al. (2011). Central chemoreceptors and neural mechanisms of cardiorespiratory control. *Brazilian Journal Medical Biology Resides, 44*(9), 883–889.

Patel, D. M. (2014). *Desmosomes of the heart: A review of clinical and mechanical analyses, Cell communication and adhesion* (pp. 109–128). London: Informa Healthcare.

Petegen, F. (2012). Ryanodine receptors: Structure and function. *Journal of Biological Chemistry., 287*(38), 31624–31632. https://doi.org/10.1074/jbc.R112.349068.

Picciotto, M. R., et al. (2002). Acetylcholine. In K. L. Davis et al. (Eds.), *Neuropharmacology: The fifth generation of progress* (pp. 3–14). Philadelphia: Lippincott Williams & Wilkins.

Pinnell, J., Turner, S., & Howell, S. (2007). Cardiac muscle physiology. *Continuing Education in Anaesthesia Critical Care & Pain, 7*(3), 85–88.

Wikibooks Contributors. (2018a). *Structural biochemistry/cell organelles/cytosol.* Wikibooks, the free textbook project. https://en.wikibooks.org/w/index.php?title=Structural_Biochemistry/Cell_Organelles/Cytosol&oldid=3365843

Wikibooks Contributors. (2018b). *Human physiology*. Wikibooks, the free textbook project. https://en.wikibooks.org/w/index.php?title=Human_Physiology&oldid=3450754

Zhilin, Q., Hu, G., Garfinkel, A., & Weiss, J. (2014). Nonlinear and stochastic dynamics in the heart. *Physics Reports, 543*, 61–162.

Chapter 8
Measured Influences of Magnetic Storms on the Heart

Abstract In this chapter we continue our investigations, described in Chaps. 5, 6, and 7, into the influences of cosmic weather dynamics (Chap. 4) on the brain and heart. In Sect. 8.1, we note that we were unable to discern visually distinct influences of geomagnetic storms on heart function, using the primary electrocardiographic (ECG), respiratory rhythm, and cardiac R-R interval data we recorded, mostly because of the large variations in these cardiac characteristics between healthy young volunteers. In our electroencephalographic (EEG) investigations, we also observed large interindividual variations in the measured data. Although our analysis of forehead–occipital theta rhythm coherence function allowed us to distinguish geomagnetic storm–influenced EEG recordings from background recordings, no such distinction was possible in our analysis of ECG investigations. We therefore analyzed measurements of the dynamics of known cardiac parameters (which are listed at the end of Sect. 8.1), and this analysis demonstrated a distinct influence of geomagnetic storms on these cardiac parameters (see Sect. 8.2).

Keywords Primary data · ECG · Cardiac parameters · R-R intervals

8.1 Measurements of the Heart in Connection with Geomagnetic Storms

This theme has been investigated in Baevsky et al. (2002), Vencloviene et al. (2013), Alabdulgader et al. (2018), Dorman (2005), Stoupel (1999), and other works. On the basis of the known cardiac characteristics (Baevsky et al. 2002), we have considered the numerical aspects of this theme along with the influence of geomagnetic storms on the brain (see References in Chap. 6).

Similarly to our findings in our electroencephalographic (EEG) investigations of the brain (see Chaps. 5 and 6), we could not visually distinguish between primary recordings made under the influence of a geomagnetic storm and those made during background geomagnetic conditions, though a number of graphics, similar to ones below, were considered (Figs. 8.1 and 8.2).

O. Novik et al., *Electromagnetic Geophysical Fields*,
https://doi.org/10.1007/978-3-319-98461-2_8

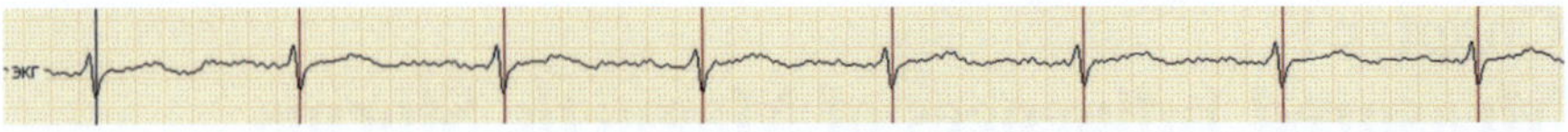

Fig. 8.1 Electrocardiographic (*ECG*) record number Strig28.01.05K3Kt, made in Moscow (see Chap. 4). The record number is structured as follows: "Strig" is the name of the examinee investigated; "28.01.05" is the date (in day–month–year order) of the recording; "K3" signifies that the geomagnetic *K*-index value was 3 at the time of the measurement (signifying the absence of a magnetic storm; see Chap. 4); and "Kt" indicates that the examinee was given a proofreading test (*PRT*; see Sect. 6.3 in Chap. 6) during the ECG, which was combined with the electroencephalographic investigation

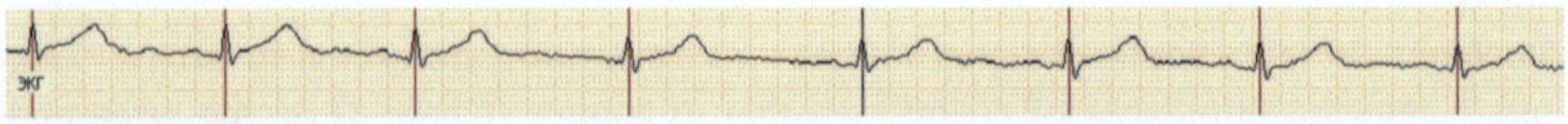

Fig. 8.2 Electrocardiographic (*ECG*) record number Mat16.06.05K5t0Kt, made in Moscow (see Chap. 4). The particularities of the record number show that the examinee was named Mat; the recording was made on June 16, 2005, during a moderate geomagnetic storm (which is denoted by "t0" in the record number), at which time the *K*-index value was 5; and the examinee was given a proofreading test during the recordings

We were unable to visually distinguish between R-R interval values recorded during quiet versus disturbed geomagnetic situations, and by the above investigations of the heart. Indeed, the R-R recordings varied greatly from one examinee to another (see the black data series in Figs. 8.3 and 8.4). Similarly, the recordings of the respiratory rhythm were not distinctive in the sense mentioned above (see the green data series in Figs. 8.3 and 8.4).

In the next section, we present tables demonstrating the changes in the parameters of heart function under the influence of a moderate geomagnetic storm ($K_p = 5$ or $K = 6$) in Moscow.

The notations for the known cardiac parameters used in our investigations of the influence of a magnetic storm on the heart are as follows (Baevsky et al. 2002):

R-R interval: Time interval (expressed in milliseconds) between two subsequent R bursts (see the discussion of R-waves in Chap. 7) on an electrocardiographic (ECG) recording.

(R-R)min: Minimal R-R time interval (expressed in milliseconds) between two subsequent R bursts on an ECG recording.

(R-R)max: Maximal R-R time interval (expressed in milliseconds) between two subsequent R bursts on an ECG recording.

RRNN: Mean duration (expressed in milliseconds) of an R-R interval.

RMSSD: Square root of the sum of squares of the differences in successive pairs of NN intervals (i.e., intervals without extrasystoles) from RR intervals) (expressed in milliseconds).

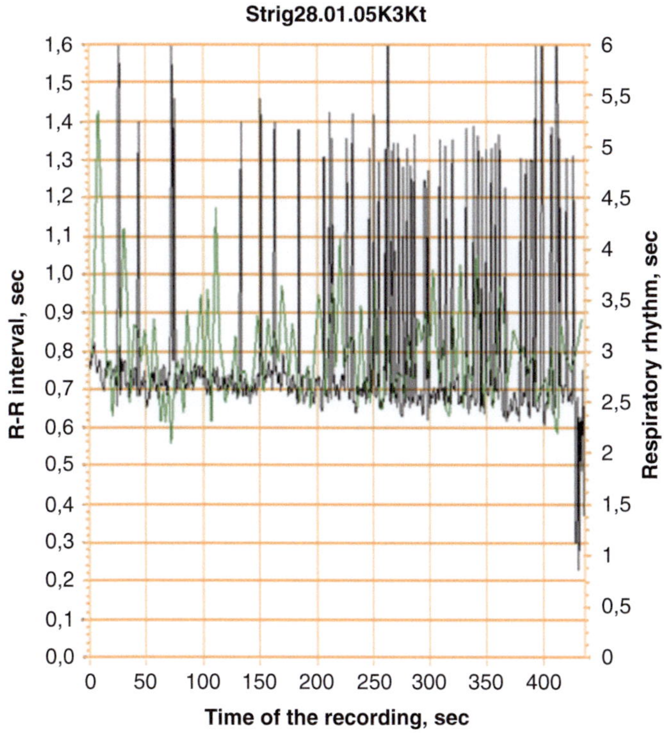

Fig. 8.3 R-R interval recording (shown in *black*) and respiratory rhythm recording (shown in *green*) in quiet (i.e., background) geomagnetic conditions, at which time the *K*-index value was 3. These recordings were made during the electroencephalographic investigations described in Chaps. 5 and 6; note the symbol "Kt" in the record number Strig28.01.05K3Kt considered here, which indicates that the examinee was given a proofreading test during the recordings

PNN50: Percentage of neighboring pairs of NN intervals during which the difference between them is more than 50 ms during the whole recording.

CV: Coefficient of variation (expressed as a percentage); $\mathrm{CV} = (\sigma/M)$ $*100$; σ is the mean square deviation (expressed in milliseconds) for the case of an R-R interval.

M: Mathematical expectation of the value of an R-R interval, i.e., the mean. $Dy = M(y - My)^2$ is the dispersion of y; the mean square deviation is $\sigma = \sqrt{Dy}$, where y is an arbitrary random value, e.g., $y =$ R-R.

LF: Duration of low-frequency oscillations of R-R intervals, expressed as a percentage of the overall duration of the recording, see the above Section HRV description.

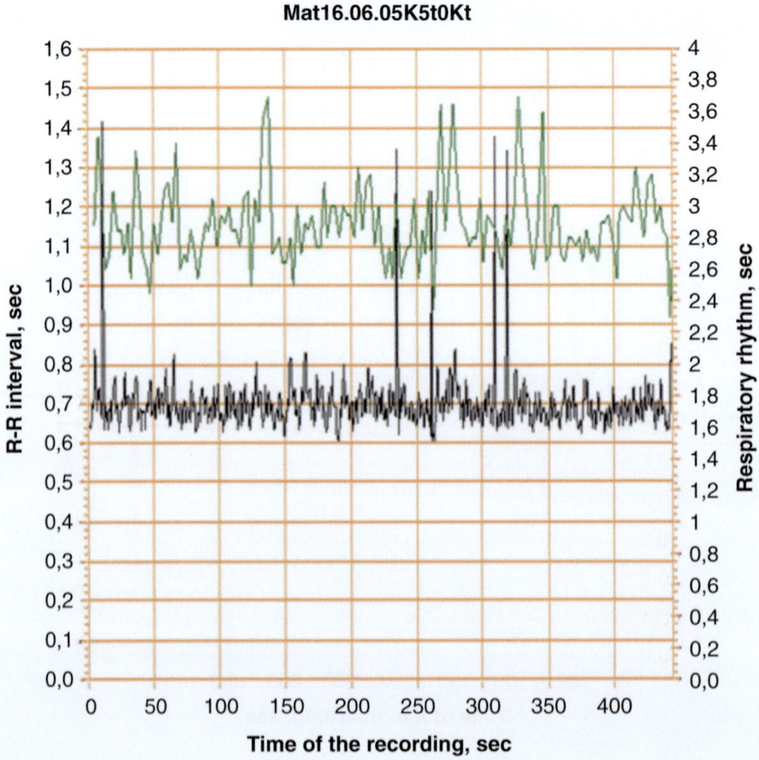

Fig. 8.4 R-R interval recording (shown in *black*) and respiratory rhythm recording (shown in *green*) during a moderate geomagnetic storm, at which time the *K*-index value was 5. These recordings were made during the electroencephalographic investigations described in Chaps. 5 and 6; note the symbol "Kt" in the record number Mat16.06.05K5t0Kt considered here, which indicates that the examinee was given a proofreading test during the recordings

HF: Duration of high-frequency oscillations of R-R intervals, expressed as a percentage of the overall duration of the recording; see the above Section HRV description.

8.2 Measurements of Changes in Cardiac Parameters

A 12-channel Polyspectr Rhythm millivoltmeter (In Russian: «Поли спектр ритм»; produced by Neurosoft, Russia) was applied by us for cardiac measurements. The signal was written in the database and then was processed with the standard program for analysis of ECGs. Processing of the signal was performed by the method for heart rate variability (HRV). The R-R interval recordings are shown in Sect. 8.1. The device used was able to register the respiratory rhythm. The data on the respiratory rhythm were recorded and processed along with the registration and processing of the ECG. The same group of examinees [students from the Third Medical Institute of Moscow (now the University)] underwent both ECG and EEG investigations at the same time. Later, at the Second Medical Institute of Moscow, EEG measurements were performed without ECGs. Before the examinees were chosen, preliminary investigations was performed, including arterial blood pressure and ECG test registration. The ECG test registration of an examinee for the investigations had to show no nontypical QRS complexes or splitting of the R peaks. Single cases of an extrasystole during 50–100 cardiac cycles were accepted. Medical drugs influencing the frequency of cardiac contractions were not used by the examinees. Before the measurements, EEG electrodes in a system with eight channels and ECG electrodes in a system of three channels were placed on each examinee. The active cardiac electrodes were placed on the left and right hands and the ground electrode on the left leg. This system of channels is considered standard for an ECG. The sensor for respiration was placed under the nose.

The results of the measurements are shown in the Tables below and illustrated by the computer graphics.

Each (R-R) number in the header row of Table 8.1 denotes the identification number assigned to an individual measurement. In this table, there are: the background values of (R-R)min and (R-R)max, [ms] (lines 1,2); the values under the influence of the geomagnetic storm (lines 3,4); and the values registered 24 h after a geomagnetic storm (lines 5,6). One can see the considerable differences in the values measured during the geomagnetic storm, as compared with the background values (Figs. 8.5, 8.6, 8.7, and 8.8; Tables 8.2, 8.3, 8.4, 8.5, and 8.6).

Table 8.1 Influence of a geomagnetic storm on the *(R-R)min* and *(R-R)max* (minimum and maximum R-R intervals); the values are expressed in milliseconds

No.	Geomagnetic conditions		(R-R)1	(R-R)2	(R-R)3	(R-R)4	(R-R)5	(R-R)6
1	Background	(R-R)min	195	230	270	780	205	720
2		(R-R)max	2135	2015	1075	1855	900	1080
3	Storm	(R-R)min	730	735	600	605	690	225
4		(R-R)max	990	1030	980	1415	2975	2620
5	t24	(R-R)min	345	185	330	155	575	
6		(R-R)max	1005	1720	1330	2030	910	

The (R-R)min and (R-R)max were influenced by the occurrence of a geomagnetic storm
t24 values registered 24 h after a storm

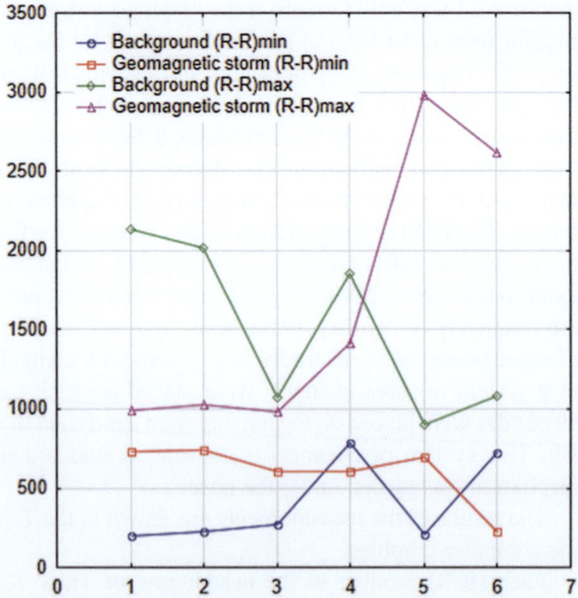

Fig. 8.5 Influence of a geomagnetic storm on cardiac characteristics. The difference (expressed in milliseconds) between the *blue line* (background) and the *red line* (geomagnetic storm) demonstrates the influence of the storm on the *(R-R)min* (the minimal R-R time interval). Two other lines demonstrate the influence of the storm on the *(R-R)max* (the maximal R-R time interval). The *horizontal axis* shows the identification numbers assigned to individual measurements

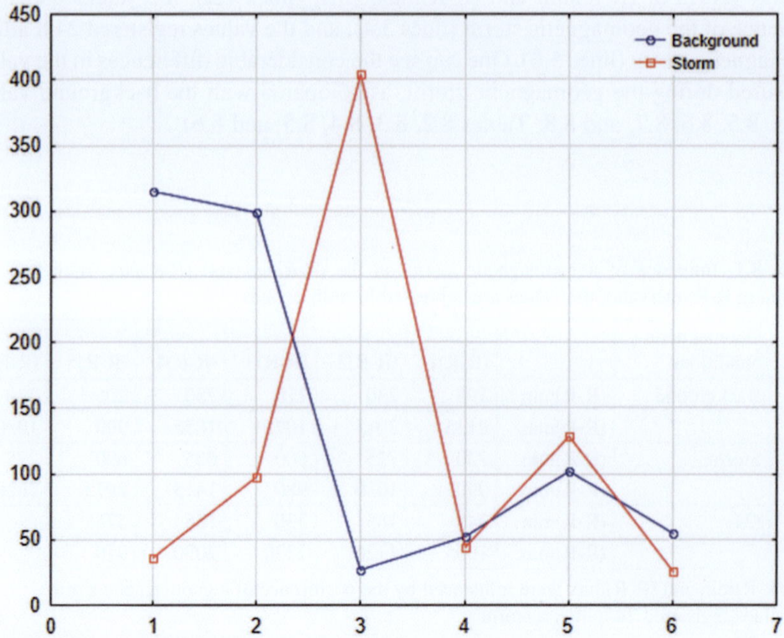

Fig. 8.6 Influence of a geomagnetic storm on the *RMSSD* (the square root of the sum of squares of the differences in successive pairs of NN intervals). One can see that the storm had a considerable influence on this cardiac characteristic. The *vertical axis* shows the time (expressed in milliseconds) and the *horizontal axis* shows the identification numbers assigned to individual measurements

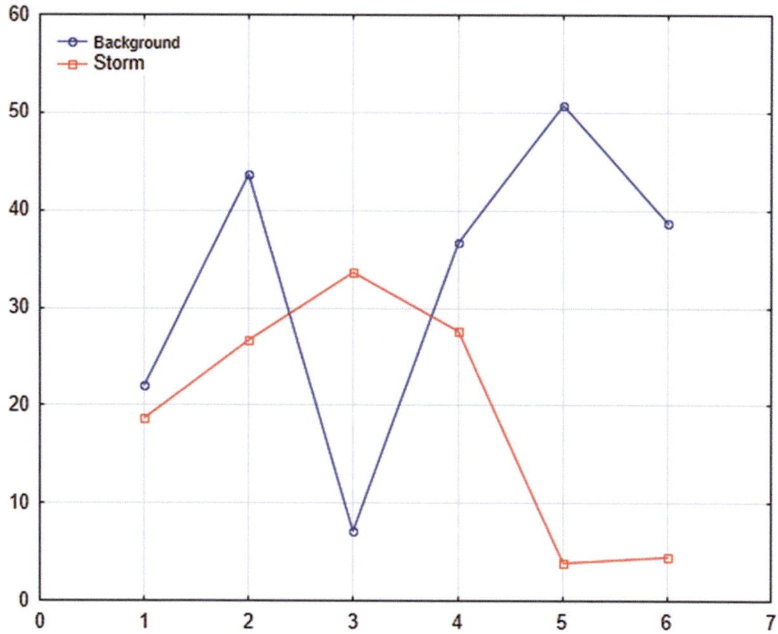

Fig. 8.7 Influence of a geomagnetic storm on the *PNN50* (the part of neighboring pairs of NN intervals during which the difference between them was more than 50 ms during the whole recording). One can see that the storm had a considerable influence. The *vertical axis* shows the PNN50 (expressed as a percentage) and the *horizontal axis* shows the identification numbers assigned to individual measurements

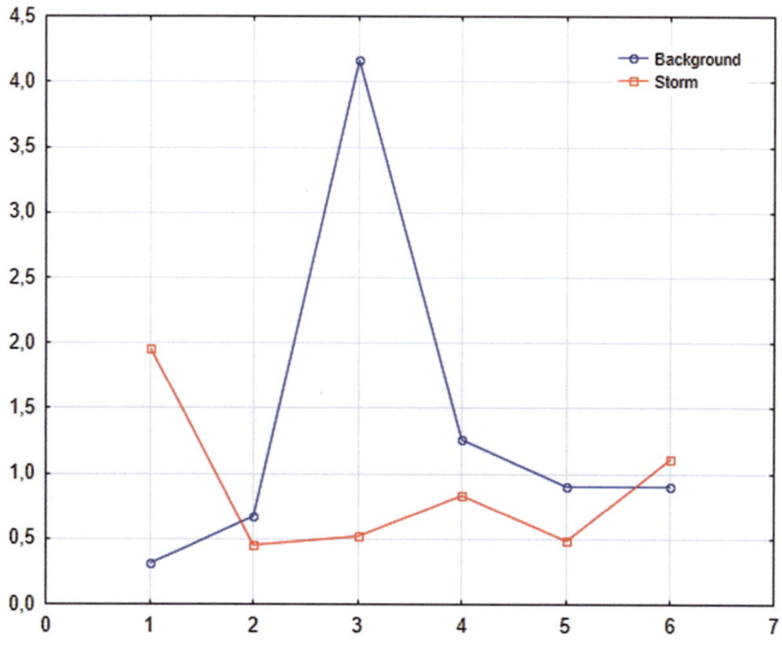

Fig. 8.8 Influence of a geomagnetic storm on the *LF/HF* (low-frequency/high-frequency) ratio

Table 8.2 Influence of a geomagnetic storm on the *RRNN* (mean R-R interval duration); the values are expressed in milliseconds

No.	Geomagnetic conditions	RRNN	RRNN1	RRNN2	RRNN3	RRNN4	RRNN5
1	Background	756	810	749	957	928	1014
2	Storm	880	697	767	752	794	819
3	t24	691	754	718	686	710	885

The RRNN parameter was not distinctly influenced by the occurrence of a geomagnetic storm
t24 values registered 24 h after a storm

Table 8.3 Influence of a geomagnetic storm on the *RMSSD* (square root of the sum of squares of the differences in successive pairs of NN intervals); the values are expressed in milliseconds

No.	Geomagnetic conditions	RMSSD	RMSSD1	RMSSD2	RMSSD3	RMSSD4	RMSSD5
1	Background	315	299	27	52	102	55
2	Storm	36	97	404	44	129	26
3	t24	136	33	42	58	97	60

The RMSSD parameter was influenced by the occurrence of a geomagnetic storm
t24 values registered 24 h after a storm

Table 8.4 Influence of a geomagnetic storm on the *PNN50* (percentage of neighboring pairs of NN intervals during which the difference between them was more than 50 ms during the whole recording)

No.	Geomagnetic conditions	(PNN50) 1	(PNN50) 2	(PNN50) 3	(PNN50) 4	(PNN50) 5	(PNN50) 6
1	Background	22	43.6	7.07	36.7	50.8	38.7
2	Storm	18.7	26.7	33.6	27.7	3.88	4.53
3	t24	27.9	2.89	27	25.1	4.37	42.1

The PNN50 was distinctly influenced by the occurrence of a geomagnetic storm
t24 values registered 24 h after a storm

Table 8.5 Influence of a geomagnetic storm on the *CV* (coefficient of variation) for the R-R interval; the values are expressed as percentages

No.	Geomagnetic conditions	CV1	CV2	CV3	CV4	CV5	CV6
1	Background	29	29.8	6.29	6.18	10.9	4.92
2	Storm	5.72	10.1	37.1	9.07	13.9	5.02
3	t24	16.3	4.73	9.18	11	12.3	6.45

The CV for the R-R interval was considerably influenced by the occurrence of a geomagnetic storm, but not as much as the other cardiac parameters analyzed in Tables 8.1, 8.3, and 8.4
t24 values registered 24 h after a storm

Table 8.6 Influence of a geomagnetic storm on changes in the cardiac *LF/HF* (low-frequency/high-frequency) ratio

No.	Geomagnetic conditions	(LF/HF)1	(LF/HF)2	(LF/HF)3	(LF/HF)4	(LF/HF)5	(LF/HF)6
1	Background	0.316	0.676	4.17	1.26	0.903	0.899
2	Storm	1.96	0.452	0.517	0.829	0.484	1.11
3	t24	0.58	1.06	0.477	0.586	0.781	0.535

The LF/HF ratio was considerably influenced by the occurrence of a geomagnetic storm
t24 values registered 24 h after a storm

8.3 Conclusion

As a finding from standard cardiological measurements performed in a group of healthy young volunteers (students at the Third Medical Higher Education Institute in Moscow) in 2005, we have shown that a number of the parameters (listed in Sect. 8.1) that are widely applied in medicine for characterization of the functioning of the heart changed distinctly under the influence of a moderate geomagnetic storm before returning to their background values within 2 days after the storm.

References

Alabdulgader, A., et al. (2018). Long-term study of heart rate variability responses to changes in the solar and geomagnetic environment. *Scientific Reports, 8*, 2663. https://doi.org/10.1038/s41598-018-20932-x. 10.01.2019

Baevsky, R. M. et al. (2002). *HRV analysis under the usage of different electrocardiography systems (methodical recommendations).* Committee of Clinic Diagnostic Apparatus and the Committee of New Medical Techniques of Ministry of Health of Russia (protocol № 4 from the 11th of April, 2002).

Dorman, L. I. (2005). Space weather and dangerous phenomena on the earth: Principles of great geomagnetic storms forecasting by online cosmic ray data. *Annales Geophysicae, 23*, 2997–3002.

Stoupel, E. (1999). Effect of geomagnetic activity on cardiovascular parameters. *Journal of Clinical and Basic Cardiology, 2*, 34–40.

Vencloviene, J., et al. (2013). The effect of solar–geomagnetic activity during and after admission on survival in patients with acute coronary syndromes. *Biometeorology, 6*, 1295–1303. https://doi.org/10.1007/s00484-013-0725-0. 10.01.2019

Chapter 9
String–Diffusion Systems of Equations

Abstract In Chaps. 9, 10, and 11 we turn our attention to the mathematical background of the geophysical results. In Chap. 9, we show that the systems of partial differential equations (PDEs) applied by us for description of the seismo-electromagnetic interaction during preparation of earthquakes, being without a type according to classical typing of differential equations in mathematical physics, are not something rare and exotic, because systems of equations with a similar structure are used in different domains of physics when nonstationary interactions of physical fields of different natures are investigated. The general structure of these systems is described in Sect. 9.1 and may be characterized as coupling of the hyperbolic and parabolic subsystems by the operators of interaction of the physical fields of different natures that are activated during the earthquake preparation process. In Sects. 9.2–9.5, we describe examples of similar systems of PDEs applied in physics (including geophysics of seismically active regions) according to different assumptions about the properties of the nonideal (from the rheological viewpoint) geological medium under consideration (e.g., anisotropy, "memory" in the case of magneto-thermo-viscoelasticity, etc.).

Keywords Typing of differential equations · String–diffusion systems of partial differential equations of precursory physics

9.1 Structure of SDSE

String–diffusion systems of equations (SDSE) are written in different forms (e.g., in the form of an equation of the third differential order; see below) and with different notations and different physical senses of known and unknown functions in the system, depending on the physical problem under consideration. But, in all cases, these systems may be written (by proper notation) in the form of two equations with the operator coefficients $\boldsymbol{H}$, $\boldsymbol{P}$, $\boldsymbol{A}$, and $\boldsymbol{B}$:

O. Novik et al., *Electromagnetic Geophysical Fields*,
https://doi.org/10.1007/978-3-319-98461-2_9

$$\begin{aligned} \boldsymbol{H}\boldsymbol{u} + \boldsymbol{A}(\boldsymbol{u}, \boldsymbol{w}) &= \boldsymbol{f} \\ \boldsymbol{P}\boldsymbol{w} + \boldsymbol{B}(\boldsymbol{w}, \boldsymbol{u}) &= \boldsymbol{\varphi} \end{aligned} \tag{9.1}$$

where the hyperbolic ($\boldsymbol{H}$) and parabolic ($\boldsymbol{P}$) systems of equations, denoted here by

$$\begin{aligned} \boldsymbol{H}\boldsymbol{u} &= \boldsymbol{f} \\ \boldsymbol{P}\boldsymbol{w} &= \boldsymbol{\varphi} \end{aligned}$$

describe the evolution of the vector fields of different natures **u** and **w** with the "sources" $\boldsymbol{f}$ and $\boldsymbol{\varphi}$ by neglect of mutual influence (i.e., interaction) of the fields; the operators $\boldsymbol{A}(\boldsymbol{u}, \boldsymbol{w})$ and $\boldsymbol{B}(\boldsymbol{w}, \boldsymbol{u})$ introduce the field interaction in the last two equations of the classical types.

But the interaction vanishes for the case of weak fields (the proper sense, i.e., metrics, of convergence $\rightarrow$ is meant):

$$\boldsymbol{A}(\boldsymbol{u}, \boldsymbol{w}) \rightarrow 0 \text{ by } \boldsymbol{w} \rightarrow 0, \boldsymbol{B}(\boldsymbol{w}, \boldsymbol{u}) \rightarrow 0 \text{ by } \boldsymbol{u} \rightarrow 0. \tag{9.2}$$

The system of equations with the operator coefficients of the form (9.1), by the above notations and the limit condition (9.2) have been named the *string–diffusion system of equations* (SDSE) by the authors because, in the simplest case, $\boldsymbol{H}$ is the operator of the one-dimensional (1D) uniform string: $\boldsymbol{H} = (\partial^2/\partial t^2 - a\partial^2/\partial x^2)$ and $\boldsymbol{P}$ is the operator of diffusion: $\boldsymbol{P} = (\partial/\partial t - b\partial^2/\partial x^2)$; here a and b are the given constants.

For example, interaction of physical fields of different natures is typical for an earthquake preparation zone with attenuation up to zero some distance away.

In this chapter we show that the systems of equations of thermo-elasticity and magneto-thermo-elasticity (MTE)—without displacement currents or with them—for different ideal and nonideal types of materials (e.g., nonuniform, anisotropic, with "memory") belong to the SDSE type. We shall not consider systems with constant coefficients (the case of a uniform material) when, with the integral transformations, the problem of determination of the SDSE type may be formulated in algebraic terms.

Interaction of conservative processes (e.g., deformation of an ideally elastic medium, electromagnetic (EM) waves without charge transfer, acoustic waves without heat transfer) with dissipative ones (e.g., heat transfer, diffusion) is described by systems of partial differential equations (PDEs), which are not of any type according to typing of PDEs in mathematical physics. The absence of a type means that the properties of the solutions of these equations, with the given physically rational additional conditions (e.g., initial and/or boundary conditions), are not known well. In particular, it is not known whether the solution exists for the time interval needed to quantitatively consider an important phenomenon, and whether the solution is unique; is its dependence upon additional data continuous? If the answer to all of these questions is positive, then the initial boundary value problem (the system of PDEs with the additional data, let us repeat) is called well posed or correctly set in the sense of Jacques Hadamard (1865–1963, France). Surely, a negative answer to any of these questions makes a system of PDEs inapplicable for the description of a physical phenomenon. For example, absence of continuous dependence of the solution upon initial and/or boundary conditions causes considerable changes in this solution by small changes in the data of the investigated problem. It is not acceptable because the data result from measurements, and these may be performed with some errors only, depending on the qualities of the devices used. The important theory of ill-posed problems for PDEs is not considered here.

It is possible that the problems with typing of the systems of PDEs of interaction of the physical fields of different natures are connected with the fact that classification of equations was created when the noninteracting fields—i.e., mechanical, thermal, EM, and gravitational—were investigated and the mutual influence of the fields of different natures was neglected though they were located in the same medium at the same time. Maybe this neglect was caused by the mathematical difficulties arising with an attempt to solve (in a classical or generalized sense) a composed system of PDEs of the interacting fields (see below). One way or another, the questions of correctness were studied mostly for systems of PDEs of known types or similar to them, and absence of a type causes difficulties with construction and investigation of the methods of solution of the initial boundary value problems and applications. The importance of investigations of the mathematical models without a type increases a great deal because the quantitative results based on them must be used for earthquake prognosis theory and practice to inform people to shut down technology (e.g., a nuclear power plant), etc.

The variety of string–diffusion (S.-D.) phenomena considered in geophysics—i.e., ones when processes of a conservative character (string) and a dissipative character (diffusion) may not be separated without considerable distortions of the fields under consideration—is too great. Correspondingly, the systems of PDEs without a type (according to classical or generalized typing) are rather different in regard to the physical sense and structure including the number of equations and their differential orders.

So, to describe the S.-D. phenomena numerically, it is necessary to determine the characteristic general properties of the systems of equations applied for description of these phenomena; to write out the generalized system of equations that has these properties and includes, as particular cases, the systems of equations of the different S.-D. phenomena; and to investigate the correctness of the initial boundary value problems for the generalized model of the S.-D. phenomena.

It is shown below that the mathematical models of the linear S.-D. phenomena may be written and investigated in the following form:

$$\begin{aligned} \boldsymbol{Hu} + \boldsymbol{Aw} &= \boldsymbol{f} \\ \boldsymbol{Pw} + \boldsymbol{Bu} &= \boldsymbol{q}. \end{aligned} \tag{9.3}$$

The string (i.e., conservative) component of the S.-D. phenomena is described by the search vector function:

$$\mathbf{u} \equiv \boldsymbol{u}(t,\boldsymbol{x}) = (u_1(t,\boldsymbol{x}), u_2(t,\boldsymbol{x}), \ldots u_{\mathrm{m}}(t,\boldsymbol{x})), \boldsymbol{x} = (x_1, x_2, \ldots x_{\mathrm{n}})$$

and the given "force" $\mathbf{f} \equiv \boldsymbol{f}(t,\boldsymbol{x}) = (f_1(t,\boldsymbol{x}), f_2(t,\boldsymbol{x}), \ldots, f_{\mathrm{m}}(t,\boldsymbol{x}))$. Neglecting dissipation, a physical phenomenon may be considered approximately as a poor conservative process. Then the system of the conservation laws controlling this process belongs to the hyperbolic type; let us denote the operator of this system by **H**:

$$\boldsymbol{Hu} = \boldsymbol{f}. \tag{9.4}$$

The dissipative component of the S.-D. phenomena is described by the search vector function

$$\mathbf{w} \equiv \boldsymbol{w}(t,\boldsymbol{x}) = (w_1(t,\boldsymbol{x}), w_2(t,\boldsymbol{x}), \ldots w_l\,(t,\boldsymbol{x}))$$

and the given "sources" $\equiv \boldsymbol{q}(t,\boldsymbol{x})$. If the influence of the conservative (string) process on the dissipative one (diffusion) is neglected, then the physical phenomenon will be described approximately by a system of equations of a parabolic type (Ladyjenskaya et al. 1967); let us denote the operator of this system by **P**:

$$\boldsymbol{Pw} = \boldsymbol{q}. \tag{9.5}$$

If the values of $\boldsymbol{u}(t, \boldsymbol{x})$ depend considerably on $\boldsymbol{w}(t, \boldsymbol{x})$, and vice versa (i.e., the conservative and dissipative components of a process $(\boldsymbol{u}(t, \boldsymbol{x}), \boldsymbol{w}(t, \boldsymbol{x}))$ are strongly interconnected), then the state of the system under consideration cannot be described by the breakable system (9.4) and (9.5). The influence of the dissipative process on the conservative one in system (9.3) is described by the operator $\boldsymbol{A}$ and the opposite influence is described by the operator $\boldsymbol{B}$.

The system of equations of the form (9.3) *with the linear operator coefficients of the types specified above shall be referred to by us as a linear SDSE.*

The described general structure (9.3) of the systems of equations of the S.-D. phenomena allows us to analyze different mathematical models of similar phenomena from the general point of view (Novik 1969).

Usually, the nonstationary magnetic (**h**) and temperature (ϑ) fields are described by parabolic equations (we suppose that the electrical conductivity value is high and the displacement current is neglected):

$$\begin{aligned} \partial_t \mathbf{h} &= \eta \Delta \mathbf{h} \\ \partial_t \vartheta &= \kappa \Delta \theta + q \end{aligned} \tag{9.6}$$

where η and κ are the coefficients of magnetic and thermal diffusion, respectively; here q is the density of the heat sources. Let us write Eq. (9.6) with the help of the parabolic operator **P** acting on the nonmechanical fields $\boldsymbol{w} = (\mathbf{h}, \vartheta)$, which have been combined because of their dissipative character, i.e., because they are described by parabolic equations, using more formal terms:

$$\mathbf{P}\mathbf{w} \equiv \frac{\partial \mathbf{w}}{\partial t} - \begin{pmatrix} \eta & 0 & 0 & 0 \\ 0 & \eta & 0 & 0 \\ 0 & 0 & \eta & 0 \\ 0 & 0 & 0 & \widetilde{\kappa} \end{pmatrix} \Delta \mathbf{w} = \begin{pmatrix} 0 \\ 0 \\ 0 \\ \mathbf{q} \end{pmatrix}. \tag{9.7}$$

We note that the deformation or motion of the medium was not included in the classical Eq. (9.7). Besides, the opposite action of the nonmechanical fields $\mathbf{w} = (\mathbf{H}, \vartheta)$ on the elastic field **u** (displacements) may not be described by the classical theory of elasticity because these fields are not considered in this theory.

Absence of interconnection between processes of conservative and dissipative characters may be described schematically by the system of Eq. (9.8):

$$\boldsymbol{Hu} = \boldsymbol{f}, \boldsymbol{Pw} = \boldsymbol{q} \tag{9.8}$$

where $\boldsymbol{H}$ is the hyperbolic operator, e.g., operator of the dynamical theory of elasticity based on the law of conservation of impulse, in this case $\boldsymbol{f} = \rho \boldsymbol{g}$, where $\boldsymbol{g}$ is acceleration of gravity and ρ is density.

Absence of a response of magnetic and temperature fields to seismic activation of a medium contradicts observations. But it will be the main property of a model if the characteristics of a nonstationary elastic field are not included in the coefficients of the nonmechanical equation to describe changes in the physical properties of a medium during preparation of an earthquake. Besides, the term of the type of a source depending on elastic field may be used in a nonmechanical equation. Surely, it is possible to use both of these possibilities in the same equation, but the only case of an additional "source"—say, $\boldsymbol{B}(\boldsymbol{u})$—will be investigated in this work for nonmechanical equations. It is necessary to take account of the opposite influence of the nonmechanical fields on the mechanical ones and to include an additional "force"—say, $\boldsymbol{A}(\boldsymbol{w})$—in the first (mechanical) equation of system (9.8).

The character or scale of action of a field of another nature **u** on the given field **w** may depend not only upon the acting field **u** but also on the state of the field under influence **w**. In other words, the "source" $\boldsymbol{B}(\boldsymbol{u})$ may depend on $\boldsymbol{w}$ and must be written in the form $\boldsymbol{B}(\boldsymbol{u}, \boldsymbol{w})$. For similar reasons, the "force" $\boldsymbol{A}(\boldsymbol{w})$, in the case of the nonmechanical processes' $\boldsymbol{w}$ influence on the mechanical ones $\boldsymbol{u}$, must be written in the form $\boldsymbol{A}(\boldsymbol{w}, \boldsymbol{u})$. Therefore, these additional terms may be considered as operators of interaction of physical fields of different natures.

So, the interdependence of geophysical fields of different physical natures, observed during field expeditions in different seismically active regions, demands that we include the interaction operators of physical fields in the above classical equations, i.e., (9.8) must be replaced by the above S.-D. system (9.1) and the limit conditions.

Equation (9.2) of weak fields for the operators $\boldsymbol{A}$ and $\boldsymbol{B}$ describes the S.-D. (i.e., between $\mathbf{u}$ and $\mathbf{w}$) interaction in a nonideal medium, e.g., an energetically saturated zone near a future epicenter area.

Conditions providing correctness of the initial boundary value problems for S.-D. systems and their discrete analogies are discussed below.

9.2 Examples of SDSE of Interacting Physical Fields

9.2.1 SDSE with a Hyperbolic Operator of the First Order

In this and the next sections, examples of S.-D. interactions will be given for different media that are of interest in geophysics. As a mathematical model, system (9.3) will be used with the hyperbolic operator of the first differential order:

$$\begin{aligned}
&\left(\frac{\partial}{\partial t}+\sum_{j=1}^{n} H_j(t,\mathbf{x})\frac{\partial}{\partial x_j}\right)\mathbf{u}(t,\mathbf{x})+\left(\sum_{j=1}^{n} A_j(t,\mathbf{x})\frac{\partial}{\partial x_j}\right)\mathbf{w}(t,\mathbf{x})=\mathbf{f}(t,\mathbf{x})\\
&\left(\frac{\partial}{\partial t}-\sum_{i,j=1}^{n} P_{ij}(t,\mathbf{x})\frac{\partial^2}{\partial x_i\partial x_j}\right)\mathbf{w}(t,\mathbf{x})+\left(\sum_{j=1}^{n} B_j(t,\mathbf{x})\frac{\partial}{\partial x_j}\right)\mathbf{u}(t,\mathbf{x})=\mathbf{q}(t,\mathbf{x}).
\end{aligned} \tag{9.9}$$

Here n is the number of coordinates, the vector functions $\boldsymbol{u}(t, \boldsymbol{x}) \in \mathrm{R}^m$ and $\boldsymbol{w}(t, \boldsymbol{x}) \in \mathrm{R}^l$ describe the string and diffusion sides of a phenomenon, respectively; $m \times m$ matrices H_j and $l \times l$ matrices P_{ij} are symmetric; and there exists a number $\nu > 0$ for which

$$\sum_{i,j=1}^{n} P_{ij}\mathbf{z}_j \cdot \mathbf{z}_i \geq \nu \sum_{i=1}^{n} \mathbf{z}_i \cdot \mathbf{z}_i, \text{by } z_i \in \mathrm{R}^l, i = 1, 2, \ldots, n$$

(i.e., the quadratic form with the matrix coefficients P_{ij} is positive definite). The scalar product of z_i and z_j is denoted by $z_i \cdot z_j$.

Let us note that similar restrictions for $m \times l$ matrices A_j and $l \times m$ matrices B_j are absent, and they may be zero if S.-D. interaction is neglected.

With these conditions, the stability and convergence of a net approximation of the solution to the Cauchy problem of system (9.3) was proofed in Novik (1969).

Let us remember that the systems of equations of the form (9.3) with the operator coefficients of the types described above will be referred to as an SDSE.

As the simplest example, let us consider the following equation of the theory of plane acoustic and magneto-hydrodynamic waves (Mikhaylov et al. 1964):

$$\frac{\partial^2 z}{\partial t^2} = B_1 \frac{\partial^2 z}{\partial x^2} + P_{11} \frac{\partial^3 z}{\partial t \partial x^2}.$$

Coefficients B_1 and P_{11} are constants, $P_{11} > 0$.
Let us apply new functions

$$u = \partial z/\partial x, v = \partial z/\partial t,$$

then this equation may be rewritten as the S.-D. system:

$$\begin{aligned} \frac{\partial u}{\partial t} - \frac{\partial v}{\partial x} = 0 \\ \frac{\partial v}{\partial t} - P_{11} \frac{\partial^2 v}{\partial x^2} - B_1 \frac{\partial u}{\partial x} = 0 \end{aligned}$$

or in the form (9.3) with the operators

$$\mathrm{A} \equiv -\partial/\partial x, \mathrm{H} \equiv \partial/\partial t, \mathrm{B} \equiv -B_1 \partial/\partial x, \mathrm{P} \equiv \partial/\partial t - P_{11} \partial^2/\partial x^2.$$

The symbol $\equiv$ is applied here and below if the authors believe that it is useful to outline that the equality is a notation or definition.

The equations of two-dimensional (2D) and three-dimensional (3D) wave motions in dissipative media of different physical natures have structure (9.3).

Let us suppose that pressure p and entropy of the unit of the mass S are the given functions of density ρ and temperature ϑ. To linearize the Navier–Stokes equation for a compressible, heat-conductive, nonviscous fluid (Landau and Lifshitz 1980, 1987), we apply the thermo-dynamical equalities

$$(\partial S/\partial \rho)_\vartheta = -\rho^2 (\partial p/\partial \vartheta)_\rho; (\partial S/\partial \vartheta)_\rho = c_V \vartheta^{-1}$$

where c_V is heat capacity at constant volume. As a result, we obtain the following system of equations for description of joint propagation of small mechanical (sound) and thermal disturbances ρ', $\boldsymbol{v}'$, and ϑ' of the density ρ_0, velocity $\boldsymbol{v}_0$, and temperature ϑ_0 of the undisturbed flow:

$$\begin{aligned} &\frac{\partial \mathbf{u}}{\partial t} + H_\alpha \frac{\partial \mathbf{u}}{\partial x_\alpha} + A_\alpha \frac{\partial \vartheta'}{\partial x_\alpha} = \mathbf{f}(t, \mathbf{x}) \\ &\frac{\partial \vartheta'}{\partial t} - \frac{\kappa}{\rho_0 c_V} \frac{\partial^2 \vartheta'}{\partial x_\alpha^2} + \left(v_{0\alpha} - \frac{1}{\rho_0 c_V} \frac{\partial \kappa}{\partial x_\alpha} \right) \frac{\partial \vartheta'}{\partial x_\alpha} + B_\alpha \frac{\partial \mathbf{u}}{\partial x_\alpha} = q(t, \mathbf{x}) \end{aligned} \tag{9.10}$$

where summation over repeated indices α, $\beta = 1, 2, 3$ is meant; κ is heat conductivity; and the vector function $\boldsymbol{u}$ combines the "conservative" variables:

$$\mathbf{u} \equiv \begin{pmatrix} c \cdot \rho'(t, \mathbf{x}) \\ \rho_0 \cdot v_1'(t, \mathbf{x}) \\ \rho_0 \cdot v_2'(t, \mathbf{x}) \\ \rho_0 \cdot v_3'(t, \mathbf{x}) \end{pmatrix}$$

$c_s = \sqrt{(\partial p/\partial \rho)_0}$ is the speed of sound; $\delta_{\alpha\beta}$ is the Kronecker delta,

$$H_\alpha \equiv \begin{pmatrix} v_{0\alpha} & c_s & 0 & 0 \\ c_s & v_{0\alpha} & 0 & 0 \\ 0 & 0 & v_{0\alpha} & 0 \\ 0 & 0 & 0 & v_{0\alpha} \end{pmatrix}, \quad A_\alpha \equiv c_s \begin{pmatrix} 0 \\ \delta_{\alpha 1} \\ \delta_{\alpha 2} \\ \delta_{\alpha 3} \end{pmatrix},$$

$$B_\alpha \equiv \frac{{c_s}^2 \vartheta_0}{\rho_0^2 c_V} (0, \delta_{\alpha 1}, \delta_{\alpha 2}, \delta_{\alpha 3}).$$

Similarly, after linearization of the Navier–Stokes (Ladyjenskaya 1969) equation for a compressible, heat-conductive, viscous fluid (Landau and Lifshitz 1987), we obtain the S.-D. system in which, contrary to (9.4), the coefficients of operator H are scalars, whereas the coefficients of operator P are matrices.

Let us consider now an electrically conductive fluid that is noncompressible and nonviscous, contrary to the previous case. We suppose that the 3D space is occupied by this fluid and that the field of its initial velocities is zero in the presence of an uniform magnetic field $\boldsymbol{h}^0 = (h_{01},\ h_{02},\ h_{03})$. We suppose that a week disturbance arises in a part of this fluid with the velocity field $\mathbf{v}' = (v_1', v_2', v_3')$.

Because of the assumptions about the fluid, propagation of the disturbance in the form of longitudinal waves and accompanying dissipation of its energy are not possible as was the case for the compressible fluid. The magneto-hydrodynamic waves arising here—namely, weak Alfven waves (Alfven 1981; Landau et al. 1984)—are not longitudinal; their velocity does not depend on the thermodynamical equation of the state of substance and the dissipation is not viscous, because it is ohmic. Though there are distinct differences between the physical natures of the weak Alfven waves and the pure acoustic ones considered above, their mathematical descriptions use close variants of system (9.3) because both types of waves are characterized by intensive interaction of dissipative and conservative processes. To show this, let us linearize the system of equations of magneto-hydrodynamics for a noncompressible, nonviscous fluid with constant electrical conductivity σ_e. We suppose that

$$\rho'(t, \boldsymbol{x}) = 0, \boldsymbol{v}_0 = 0$$

and write the magnetic field intensity and pressure in the respective forms

$$\boldsymbol{h}(t,\boldsymbol{x}) = \boldsymbol{h}^0 + \boldsymbol{h}'(t,\boldsymbol{x}), p(t,\boldsymbol{x}) = p_0 + p'(t,\boldsymbol{x})$$

where $\rho'(t, \boldsymbol{x})$, $p'(t, \boldsymbol{x})$, and $\boldsymbol{h}'(t, \boldsymbol{x}) = \left(h'_1, h'_2, h'_3\right)$ are small disturbances.

We take account of div $\boldsymbol{v} = 0$ and div $\boldsymbol{h} = 0$ and apply *div* to all terms of the linearized equation of motion. Then we have $p' + \frac{1}{4\pi} \; h_{0\alpha} h'_\alpha$, is a harmonic function everywhere in the 3D space. The small disturbances are bounded uniformly with respect to $\boldsymbol{x}$ and then, according to the theorem of Liouville, this function is the integral of motion of the linearized system. So the linearized equations of motion of the fluid under consideration may be written in the form (magnetic permeability is supposed to be uniform and equal to 1):

$$\frac{\partial v'_j}{\partial t} - \frac{1}{4\pi\rho_0} h_{0\alpha} \frac{\partial h'_j}{\partial x_\alpha} = 0$$

$$\frac{\partial h'_j}{\partial t} - \frac{1}{4\pi\sigma_e} \frac{\partial^2 h'_j}{\partial x_\alpha^2} - h_{0\alpha} \frac{\partial v'_j}{\partial x_\alpha} = 0.$$

Here $j = 1, 2, 3$ and summation over the repeated index $\alpha = 1, 2, 3$ is meant.

Linearization of equations of magneto-hydrodynamics for a compressible fluid with at least one nonzero coefficient κ, $1/\sigma_e$, or η (viscosity) leads to an S.-D. of type (9.3).

S.-D. systems of linear equations of type (9.3) are used in the theory of finite amplitude waves (i.e., not only for the above case of a small amplitude) by application of the perturbation method for description of attenuation of these waves.

Besides the wave S.-D. processes, there are models of flow of water or solutions in capillary-porous bodies and other phenomena of heat–mass transfer when the patterns of the diffusion type are combined with the hydrodynamic conservation laws.

9.2.2 SDSE of Thermo-viscoelasticity as a Possible Mathematical Model of Thermal Evolution of Active Lithosphere Zones

Analysis of the geodynamical processes in the lithosphere is not possible without consideration of the wide spectrum of the mechanical states of its different regions, including regimes of the hydrodynamic type in lower floors and elastic behavior with anisotropy up to destruction (microcracks—main cracks) in the upper lithospheric floors (Kashinzev 2001).

For quantitative description of the evolution of the lithospheric fields, it is necessary to use data on the geological history of the region and take account of the “memory” of the mechanically nonideal material of the geological medium.

To get the numerical characteristics of the phenomenon of "memory" is a difficult physical problem.

The rheological models of the dynamic thermo-viscoelasticity theory of nonuniform, anisotropic media have the properties mentioned here, including variety of the possible regimes of motion. Therefore, similar models may be applied for investigations of evolution of the lithosphere, geodynamics, and interaction of the fields of different natures, along with the more traditional Navier–Stokes equation (Novik 1988).

The principal mathematical difficulty by this approach is that the *systems of integro-differential equations with partial derivatives include operators of different mathematical types* due to the different physical natures of the fields interacting over geological time. As a result, these systems of equations are not of any type that has been investigated well enough in mathematical physics, i.e., the situation here is similar to that in Sect. 9.1, but the problem of the "memory" of the geological medium has been added.

So, the system of equations of the dynamic thermo-viscoelasticity theory (Christensen 1982; Lakes 2009) may be represented by the form (9.3):

$$\mathrm{H}\boldsymbol{u} + A\vartheta = \boldsymbol{f}$$

$$P\vartheta + \mathrm{B}\boldsymbol{u} = \mathrm{q}$$

where $\vartheta(t, \boldsymbol{x})$ denotes temperature, $\boldsymbol{u}(t, \boldsymbol{x}) = (u_1, u_2, u_3)$ denotes displacements determining the deformation of a layer, $\boldsymbol{x} = (x_1, x_2, x_3)$ denotes coordinates, and t denotes time. The operators H, P, A, and B depend on evolution of fields $\boldsymbol{u}$ and ϑ during the time from the beginning $t = 0$ of the studied period up to the current moment $t \leq T$; T is the duration of the period:

$$(\mathbf{Hu})_i = \rho(t)\frac{\partial^2 u_i}{\partial t^2} - \frac{\partial}{\partial x_j}\int_0^t G_{ijkl}(t,t')\frac{\partial \varepsilon_{kl}}{\partial t'}(t')dt'$$

$$\mathrm{P}\vartheta = \frac{\partial}{\partial t}\int_0^t m(t,t')\frac{\partial \vartheta}{\partial t'}(t')dt' - \frac{\partial}{\partial x_j}\left(k_{ij}(t)\frac{\partial \vartheta}{\partial x_i}(t)\right)$$

$$(\mathrm{A}\vartheta)_i = \frac{\partial}{\partial x_j}\int_0^t \phi_{ij}(t,t')\frac{\partial \vartheta}{\partial t'}(t')dt'$$

$$\mathbf{Bu} = \vartheta_0\frac{\partial}{\partial t}\int_0^t \phi_{ij}(t,t')\frac{\partial \varepsilon_{ij}}{\partial t'}(t')dt'.$$

In these expressions the vector of coordinates $\boldsymbol{x}$ is the parameter, all values are considered at the same point $\boldsymbol{x}$ (which has been omitted for brevity), ϑ_0 is the temperature of the readout, the kernels of the integral operators (known as kernels of relaxation in this context) describe the dependence of recent states of the interacting fields on their history, and summation over repeated indices $j, k, l = 1, 2, 3$ is meant.

Computations of interaction of nonstationary fields of different natures in a nonuniform 3D medium are possible only on the basis of careful justification of the algorithms applied. It is necessary because we need to be sure that the calculated quantitative particularities of the evolution of the fields are determined by the physical assumptions about the medium and processes, and may not be regarded as the result of accumulation of errors of the ultimate computer arithmetic during the long working (because the geological scale of time is considered) of a composed algorithm by absence of the possibility to use the experience of calculations of the separate (noninteracting) fields.

Moving ahead, we note that correctness (numerical stability) will be proved for the discretized initial boundary value problem of the system of equations of thermo-viscoelasticity. In particular, it will be demonstrated that the sum of mechanical and thermal energy is bounded at any moment of time by some value determined by the initial state of the medium under consideration, its physical characteristics, external loads of different natures, the mass force, and internal sources of energy.

To make sure that the system of equations of thermo-viscoelasticity may be written in the form of the S.-D. system (with the symmetric matrices by the second-order derivatives in the hyperbolic operator), let us transform the mechanical part of the stress tensor:

$$\begin{aligned}\sigma_{ij}^{M} &= \int_0^t G_{ijkl}(t,t')\frac{\partial \varepsilon_{kl}}{\partial t'}(t')dt' \\ &= \int_0^t \left\{\frac{\partial}{\partial t'}(G_{ijkl}(t,t')\varepsilon_{kl}(t')) - \varepsilon_{kl}(t')\frac{\partial}{\partial t'}G_{ijkl}(t,t')\right\}dt' \\ &= G_{ijkl}(t,t)\varepsilon_{kl}(t) - G_{ijkl}(t,0)\varepsilon_{kl}(0) \\ &\quad - \int_0^t \varepsilon_{kl}(t')\frac{\partial}{\partial t'}G_{ijkl}(t,t')dt'.\end{aligned}$$

Then

$$\begin{aligned}\frac{\partial}{\partial x_j}\sigma_{ij}^{M} &= G_{ijkl}(t,t)\frac{\partial^2 u_k}{\partial x_l \partial x_j}(t) + \varepsilon_{kl}(t)\frac{\partial}{\partial x_j}G_{ijkl}(t,t) \\ &\quad - \frac{\partial}{\partial x_j}(G_{ijkl}(t,0)\varepsilon_{kl}(0)) - \frac{\partial}{\partial x_j}\int_0^t \varepsilon_{kl}(t')\frac{\partial}{\partial t'}G_{ijkl}(t,t')dt'\end{aligned}$$

because

$$\begin{aligned}G_{ijkl}(t,t)\varepsilon_{kl}(t) &= \frac{1}{2}G_{ijkl}(t,t)(u_{k,l}(t) + u_{l,k}(t)) \\ &= \frac{1}{2}G_{ijkl}(t,t)u_{k,l}(t) + \frac{1}{2}G_{ijlk}u_{l,k}(t) \\ &= G_{ijkl}(t)u_{k,l}(t)\end{aligned}$$

because of the symmetry of the kernel of relaxation $G_{ijkl} = G_{ijlk}$ and the form of the tensor of deformation $\varepsilon_{kl} = ½(u_{k,l} + u_{l,k})$. Here and below we use the short notation for derivatives:

$$f,_i \equiv \partial f / \partial x_i.$$

The terms with the second-order derivatives in $\partial\sigma_{ij}^M / \partial x_i \equiv \sigma_{ij,i}^M$ may be written in the matrix form:

$$\tilde{H}^{lj}\mathbf{u}_{,lj} + \tilde{H}^{jl}\mathbf{u}_{,jl} = \frac{1}{2}\tilde{H}^{lj}(\mathbf{u}_{,jl} + \mathbf{u}_{,lj}) + \frac{1}{2}\tilde{H}^{jl}(\mathbf{u}_{,jl} + \mathbf{u}_{,lj})$$

with the nonsymmetric on indices l, j matrices (l, j is the number of a matrix):

$$\left(\tilde{H}^{lj}\right)_{ik} = G_{ijkl}, \quad \left(\tilde{H}^{jl}\right)_{ik} = G_{ilkj}$$

where i, $k = 1, 2, 3$. The upper indices determine the number of a matrix and the lower ones determine its element.

But the expression under consideration may be rewritten in the form

$$\tilde{H}^{lj}\mathbf{u}_{,lj} + \tilde{H}^{jl}\mathbf{u}_{,jl} = \frac{1}{2}\left(\tilde{H}^{jl} + \tilde{H}^{lj}\right)(\mathbf{u}_{,jl} + \mathbf{u}_{,lj}) = H^{jl}\mathbf{u}_{,jl} + H^{lj}\mathbf{u}_{,lj}$$

where $H^{lj} = H^{jl} \equiv \frac{1}{2}\left(\tilde{H}^{lj} + \tilde{H}^{jl}\right)$ are symmetric matrices and j, $l = 1, 2, 3$. Indeed, the element ik (the first index is the number of a line, the second one is the number of a column) of the matrix H^{lj} has the form $½(G_{ijkl} + G_{ilkj})$; therefore, the symmetric element for it is

$$½\left(G_{kjil} + G_{klij}\right) = ½\left(G_{ilkj} + G_{ijkl}\right)$$

and one can see that the symmetric elements are equal. Here we have used the known symmetry of the kernels of relaxation (Christensen 1982) by transposition of the first and second pairs of indices:

$$G_{ijkl} = G_{klij}.$$

So, the system of equations of the theory of thermo-viscoelasticity may be written as the S.-D. system after appropriate choice of its matrix coefficients:

$$\left(\partial_t^2 - H_{ij}(t,\mathbf{x})\partial_{ij}^2 + H_{0i}(t,\mathbf{x})\partial_{ti}^2\right)\mathbf{u}(t,\mathbf{x}) + A_i(t,\mathbf{x})\partial_i\mathbf{w}(t,\mathbf{x})$$
$$+\int_0^t \left(\mathrm{H}_{ij}(t,t',\mathbf{x})\partial_{ij}^2\mathbf{u}(t',\mathbf{x}) + \mathrm{A}_{ij}(t,t',\mathbf{x})\partial_{ij}^2\mathbf{w}(t',\mathbf{x})\right)dt'$$
$$+ l_1(\mathbf{u},\mathbf{w}) = f(t,\mathbf{x})$$
$$\left(\partial_t - P_{ij}(t,\mathbf{x})\partial_{ij}^2\right)\mathbf{w}(t,\mathbf{x}) + \left(B_{ij}(t,\mathbf{x})\partial_{ij}^2 + B_{0i}(t,\mathbf{x})\partial_{ti}^2\right)\mathbf{u}(t,\mathbf{x})$$
$$+\int_0^t \left(\mathrm{P}_{ij}(t,t',\mathbf{x})\partial_{ij}^2\mathbf{w}(t',\mathbf{x}) + \mathrm{B}_{ij}(t,t',\mathbf{x})\partial_{ij}^2\mathbf{u}(t',\mathbf{x})\right)dt'$$
$$+ l_2(\mathbf{u},\mathbf{w}) = q(t,\mathbf{x})$$

where

$$\partial_i \equiv \frac{\partial}{\partial x_i}, \quad \partial_{ij}^2 \equiv \frac{\partial^2}{\partial x_i \partial x_j}, \quad \partial_{tj}^2 \equiv \frac{\partial^2}{\partial t \partial x_i}$$

and the notations used were

$$\boldsymbol{x} = (x_1, \ldots, x_n); \boldsymbol{u} = (u_1, \ldots, u_m); \boldsymbol{w} = (w_1, \ldots, w_l)$$

where n, m, l may be equal to 1, 2 or 3; summation over repeated indices is meant from 1 up to n. Coefficients of the operators H, P, A, and B are denoted by the same letters with indices and they are the matrices with dimensions relevant to vectors $\boldsymbol{u}$, $\boldsymbol{w}$. The operators with the derivatives of the first order are denoted there by l_1 and l_2 (terms without derivatives of the unknown functions have been omitted because they are not needed here to determine the type of the system of PDEs):

$$l_1(\mathbf{u},\mathbf{w}) = \left(H_0(t,\mathbf{x})\partial_t + H_i(t,\mathbf{x})\partial_i\right)u(t,\mathbf{x})$$
$$+\int_0^t \left(\mathrm{H}_i(t,t',\mathbf{x})\partial_i\mathbf{u}(t',\mathbf{x}) + \mathrm{A}_i(t,t',\mathbf{x})\partial_i\mathbf{w}(t',\mathbf{x})\right)dt'$$
$$l_2(\mathbf{u},\mathbf{w}) = P_i(t,\mathbf{x})\partial_i\mathbf{w}(t,\mathbf{x}) + \left(B_0(t,\mathbf{x})\partial_t + B_i(t,\mathbf{x})\partial_i\right)\mathbf{u}(t,\mathbf{x})$$
$$+\int_0^t \left(\mathrm{P}_i(t,t',\mathbf{w})\partial_i\mathbf{w}(t',\mathbf{x}) + \mathrm{B}_i(t,t',\mathbf{x})\partial_i\mathbf{u}(t',\mathbf{x})\right)dt'.$$

The coefficients of this S.-D. system are determined by the kernels of relaxation and their derivatives of no higher than the second order with respect to x_i and t.

By determination of the kernels of relaxation according to the thermo-baric conditions, the system of equations of the coupled dynamic theory of thermo-viscoelasticity (Lokshin and Suvorova 1982) can be applied for investigation of attenuation of seismic waves in different lithospheric floors.

The system of equations of the coupled dynamic theory of thermo-elasticity for nonuniform, anisotropic media, which is the partial case of the considered thermo-viscoelasticity S.-D. system for the case of vanishing memory, describes attenuation of seismic waves in the crust (Novik 1983a, b; Novik et al. 2000) when problems of seismic prospecting or acoustic logging of boreholes are concerned.

9.3 SDSE of Magneto-thermo-viscoelasticity as a Mathematical Model of Seismic Disturbances of EM and Thermal Lithosphere Fields with Low-Resistivity Structures

The MTE theory (Hutter and Ven 1978; Maugin 1988)—which may be regarded approximately as a synthesis of the theories of elasticity, the electrodynamics of moving media, and nonequilibrium thermodynamics—has been developed mostly during the last few decades. Let us note that this has also been a time of considerable progress of magneto-telluric investigations into the electrical conductivity of seismically active regions. Geophysical investigations of magneto-elastic waves have been previously published (Knopoff 1955; Keylis-Borok and Monin 1959).

So, let us concretize S.-D. system (9.7) and (9.8) to get the form applicable to the problem of interaction of geophysical fields.

Thermo-mechanical interaction will be considered in the framework of the linear theory of thermo-viscoelasticity (Christensen 1982; Iesan and Quintanilla 2013; Lakes 2009). According to this theory, the stress at the current time t depends on the history of deformation in the previous time τ, but this dependence weakens with increasing $t - \tau$ (the hypothesis of decaying memory of a medium). The system of equations of the linear theory of thermo-viscoelasticity (see the above references) may be written for the case of a nonuniform medium as the following S.-D. system:

$$\begin{aligned} \mathrm{H}\mathbf{u} + \mathrm{A}_{TD}(\mathbf{u}, \vartheta) &= \rho\mathbf{g} \\ \mathrm{P}_T\vartheta + \mathrm{B}_{DT}(\mathbf{u}, \vartheta) &= q. \end{aligned} \tag{9.11}$$

Here H is the operator of the dynamical theory of viscoelasticity:

$$\mathrm{H}\boldsymbol{u} \equiv \rho\frac{\partial^2 \boldsymbol{u}}{\partial t^2} - Div\widehat{\sigma} \tag{9.12}$$

where $\boldsymbol{u} = (u_1, u_2, u_3)$ is the field of elastic displacements, ρ is the density, the ith component $\left[Div\widehat{\sigma}\right]_i$ of the vector $Div\ \widehat{\sigma}$ (i.e., divergence of the stress tensor) is determined by the sum $\left[Div\widehat{\sigma}\right]_i \equiv \partial\sigma_{i\alpha}{}^{i\alpha}/\partial x_\alpha$, and the tensor of the viscoelastic stresses

$$\sigma_{ij} \equiv \int_0^t G_{ijkl}(t,t') \frac{\partial \varepsilon_{kl}}{\partial t'}(t')dt'$$

(there is summation over repeated indices $k, l = 1, 2, 3$) is determined by the Cauchy tensor of deformation

$$\varepsilon_{ij} \equiv \frac{1}{2}\left(u_{i,j} + u_{j,i}\right) \tag{9.13}$$

and the tensor of relaxation G_{ijkl} characterizing the viscoelastic properties of the medium.

So, the rheological strain–stress relation with the history of deformations, described by the Cauchy tensor (9.13), are considered. In other words, small displacements are under consideration here, and physical and geometric nonlinearities are neglected for the case of a medium with memory. It is possible that similar restrictions can be applied in some models of the *precursory* physics (Chaps.1–3 and 9).

The operator

$$\mathrm{A}_{TD}(\boldsymbol{u}, \vartheta) = \mathrm{A}_{TD}(\vartheta) = -Div\widehat{\sigma}_T$$

describes the influence of the temperature field on the deformation process; $\widehat{\sigma}_T$ is the tensor of thermo-stresses included in the tensor of thermo-viscoelastic (*TV*) stresses $\widehat{\sigma}_{TV} = \widehat{\sigma} + \widehat{\sigma}_T$, and $\widehat{\sigma}_T$ depends on the history of the temperature field:

$$\left(\widehat{\sigma}_T(\vartheta)\right)_{ij} = -\int_0^t \phi_{ij}(t,\tau) \frac{\partial \vartheta}{\partial \tau}(\tau)d\tau \tag{9.14}$$

where ϑ is a small deviation of temperature from the reference temperature ϑ_0.

In system (9.11), the heat conduction operator P_T has the form

$$\mathrm{P}_T\vartheta \equiv \frac{\partial}{\partial t}\int_0^t m(t,\tau)\frac{\partial \vartheta}{\partial \tau}(\tau)d\tau - \mathrm{div}(\kappa \;\mathrm{grad}\; \vartheta). \tag{9.15}$$

Here m is the thermal function of relaxation of the medium with memory and κ is the coefficient of thermal conductivity. Heat generation during passage of the current through the contact of different materials is neglected, as well as the addition (it is proportional to the density of the current) to the heat flow κ grad ϑ; see Maugin (1988) about the Peltier effect.

In the model considered now, the operator $\mathrm{B}_{DT}(\boldsymbol{u}, \vartheta)$ of the influence of deformations on the temperature field is independent from ϑ:

$$\mathrm{B}_{DT}(\mathbf{u}) \equiv \vartheta_0 \frac{\partial}{\partial t} \int_0^t \phi_{ij}(t,\tau) \frac{\partial \varepsilon_{ij}}{\partial \tau}(\tau) d\tau \tag{9.16}$$

where summation over repeated indices is meant from 1 to 3. The spatial dependence of the characteristics of the medium

$$\rho, G, \varphi, m, \kappa, \sigma_{\mathrm{e}}, \pi_0, \varepsilon_{\mathrm{e}}, \mu_{\mathrm{e}}$$

acceleration of gravity $\boldsymbol{g}$, and the density of the heat sources q are omitted for brevity. We suppose that $\boldsymbol{g}$ is independent from time t during seismic destabilization of nonuniform media. Let us note that the data on seismic changes of the gravitational field [the variations of $\boldsymbol{g}$ are about 10^{-6} m/s^2 or less (Mei Shirong 1992)] are not sufficient for calculation of interactions of geophysical fields. But the preliminary estimations, based on the data about seismic disturbances of the force of gravity, show that seismic variations of the gravity field are not able to considerably change the results of the calculations that will be demonstrated below. The problem of gravitational–mechanical–EM–thermal interaction is not considered in this work, but the stationary (in the field of gravity) state of the elastic field of a geological block is considered as one of the possible initial conditions before the arrival of the seismic wave.

Moving over *from the thermo-mechanical* interaction *to the thermo-mechanical-EM* one, we describe the interactions of the EM field with the nonstationary viscoelastic and temperature fields. In other words, the force created by the EM field must be included in the equation of motion [the first equation in system (9.11)]; analogically, the source depending on the EM field must be included in the equation of heat conductivity [the second equation in system (9.11)]. The opposite influence of the mechanical and thermal fields on the EM field must be considered in accordance with the principles of the electrodynamics of moving media and included in the S.-D. system of the thermal–mechanical–EM interaction in the lithosphere domain with conductive structures (Tamm 1979; Landau et al. 1984; Landau and Lifshitz 1975; 1986).

Let us denote the laboratory frame (i.e., the reference system) by K. Let us denote by K' the local inertial frame moving, at an arbitrarily fixed moment t' and at an arbitrarily fixed point $\boldsymbol{x}'$, along with the infinitesimal volume of the medium. By going over from K to K', the field values are transformed according to the Lorentz transformation for electrical and magnetic fields (Landau and Lifshitz 1975; Feynman et al. 2005):

$$\boldsymbol{E}' = \boldsymbol{E} + \boldsymbol{v} \times \boldsymbol{B}, \boldsymbol{D}' = \boldsymbol{D}, \boldsymbol{H}' = \boldsymbol{H} - \boldsymbol{v} \times \boldsymbol{D}, \boldsymbol{B}' = \boldsymbol{B}, \boldsymbol{j}' = \boldsymbol{j} - \rho_{\mathrm{e}} \boldsymbol{v}. \tag{9.17}$$

Here the dashed letters denote the field values in K'. In frame K

$$\boldsymbol{E} = \boldsymbol{E}(t,\boldsymbol{x}), \boldsymbol{B} = \boldsymbol{B}(t,\boldsymbol{x}), \boldsymbol{D} = \boldsymbol{D}(t,\boldsymbol{x}), \boldsymbol{H} = \boldsymbol{H}(t,\boldsymbol{x}), \boldsymbol{j} = \boldsymbol{j}(t,\boldsymbol{x}), \rho_{\mathrm{e}} = \rho_{\mathrm{e}}(t,\boldsymbol{x})$$

are the values of the intensity of the electrical field, magnetic induction, electrical induction, magnetic field intensity, electrical current density, and electrical charge density, respectively; $(t, \boldsymbol{x})$ denotes temporal and spatial coordinates corresponding to $(t', \boldsymbol{x}')$ according to the Lorentz transformation; and $\boldsymbol{v}$ is the velocity of point $\boldsymbol{x}'$ at the moment t in K. As a matter of fact, transformations (9.17) are the known approximations of the Lorentz transformations for the special case of a *slowly moving medium* (c is the light velocity in a vacuum):

$$(v/c)^2 << 1.$$

This supposition is natural for our case of a medium moving as a result of seismic deformations.

The point $(t', \boldsymbol{x}')$ is resting in frame K'; therefore, the motion may be neglected in an infinitesimal vicinity of $(t', \boldsymbol{x}')$ and Ohm's law may be written in the form

$$\boldsymbol{j}' = \sigma_{\mathrm{e}}\boldsymbol{E}' - \pi_0 \cdot \mathrm{grad}'\vartheta$$

where grad′ denotes differentiation with respect to $\boldsymbol{x}'$, π_0 is the coefficient of the thermo-electric current, and σ_{e} is the electrical conductivity of the medium (let us remember that the stress tensor is denoted by $\widehat{\sigma}$ and its components are σ_{ij}).

Using semiclassical transformation (9.17), we get Ohm's law in K:

$$\boldsymbol{j} = \sigma_{\mathrm{e}}(\boldsymbol{E} + \boldsymbol{v} \times \boldsymbol{B}) + \rho_{\mathrm{e}}\boldsymbol{v} - \pi_0\,\mathrm{grad}\vartheta. \tag{9.18}$$

Because of the supposition about a *slowly moving medium*, the scale in K' and in K is the same; therefore, grad′ may be replaced by grad.

Now let us consider Maxwell's equation in K:

$$\mathrm{rot}\,\boldsymbol{H} = \boldsymbol{j} + \partial \boldsymbol{D}/\partial t.$$

The induction $\boldsymbol{D}$ of the electrical field is supposed to be proportional to its intensity $\boldsymbol{D} = \varepsilon_{\mathrm{e}}\boldsymbol{E}$, where $\varepsilon_{\mathrm{e}} = \varepsilon_0\varepsilon_r$ is the dielectric permeability, ε_r is the relative dielectric permeability (it is not more than 10 for most of the rocks), and $\varepsilon_0 = 8.854 \cdot 10^{-12}$ F/m (SI units are used) is the dielectric permeability of a vacuum.

The displacement current $\partial \boldsymbol{D}/\partial t$ may be estimated as follows:

$$\partial \boldsymbol{D}/\partial t \propto f \cdot \mid \varepsilon_{\mathrm{e}} \mid \cdot E$$

where f is the typical frequency, i.e., one of the order of 1 Hz for the geophysical fields considered below. Therefore, the displacement current is negligible as compared with the conductivity current $\sigma_{\mathrm{e}}\boldsymbol{E}$ because the electrical conductivity of the rocks considered below is more than 0.0001 S/m. Therefore, we get approximately

$$\operatorname{rot} \boldsymbol{H} = \boldsymbol{j} \tag{9.19}$$

and $\operatorname{div} \boldsymbol{j} = 0$. Then from the law of conservation of charge $\partial\rho_e/\partial t + \operatorname{div} \boldsymbol{j} = 0$, we have it that ρ_e is independent from time. Let us believe that *the medium was neutral initially*, i.e., $\rho_e = 0$ everywhere for $t = 0$, and then this characteristics of the medium will be preserved:

$$\rho_e = 0, \text{everywhere for } t > 0.$$

Now let us substitute $\boldsymbol{j}$ from (9.18) by $\rho_e = 0$ into (9.19) and multiple the result by σ_e^{-1}, then we apply *rot* term by term:

$$\operatorname{rot}\left(\sigma_e^{-1}\operatorname{rot}\mathbf{H}\right) = \operatorname{rot}\left(\sigma_e^{-1}\mathbf{j}\right) = \operatorname{rot}\mathbf{E} + \operatorname{rot}[\mathbf{v} \times \mathbf{B}] - \operatorname{rot}\left(\sigma_e^{-1}\pi_0 \operatorname{grad} \vartheta\right).$$

Using another Maxwell's equation (the generalized law of Faraday),

$$\operatorname{rot} \boldsymbol{E} = -\partial \boldsymbol{B}/\partial t,$$

we receive finally the equation describing evolution of the magnetic field:

$$\mathrm{P}_M(\boldsymbol{w}) + \mathrm{B}_{DTM}(\boldsymbol{u}, \boldsymbol{w}) = 0 \tag{9.20}$$

where $\boldsymbol{w} = (\boldsymbol{H}, \vartheta)$ combines nonmechanical fields, $\boldsymbol{u}$ is the vector field of mechanical displacements, $\boldsymbol{v}$ is the field of velocity, $\partial \boldsymbol{u}/\partial t = \boldsymbol{v}$, and the operators P_M and B_{DTM} are defined as follows:

$$\begin{aligned} \mathrm{P}_M(\mathbf{w}) &\equiv \frac{\partial \mathbf{B}}{\partial t} + \operatorname{rot}\left(\sigma_e^{-1}\operatorname{rot}\mathbf{H}\right) \\ \mathrm{B}_{DTM}(\mathbf{u}, \mathbf{w}) &\equiv \operatorname{rot}\left(\sigma_e^{-1}\pi_0 \operatorname{grad} \vartheta\right) - \operatorname{rot}[\mathbf{v} \times \mathbf{B}]. \end{aligned} \tag{9.21}$$

The operator $\mathrm{B}_{DTM}(\boldsymbol{u}, \boldsymbol{w})$ describes the influence of the deformation and thermal fields on the magnetic field. The magnetic field intensity $\boldsymbol{H}$ and temperature ϑ are being considered in diffusion approximation [an approach to the theory of heat transfer without the paradox of infinite velocity of heat propagation can be found in Shashkov et al. (2004)].

The impact of an EM field on the fixed infinitesimal volume ΔV will be taken into account as the force acting on charges in this volume in the frame K. The volume density of this EM force, along with the values of the EM field, depend on the frame and must be added to $\rho \boldsymbol{g}$, i.e., included in the classical equation of motion.

First of all, let us consider the relativistic equation of motion of charge e in the EM field (Landau and Lifshitz 1975):

$$\partial \boldsymbol{p}/\partial t = e(\boldsymbol{E} + \mu_0 \boldsymbol{v} \times \boldsymbol{H})$$

where $\mathbf{p} = m_e \mathbf{v}/\sqrt{1-(v/c)^2}$ is the impulse of the particle, m_e is its rest mass, and μ_0 is the magnetic permeability of a vacuum. The intensities of the electrical and magnetic fields $\boldsymbol{E}$ and $\boldsymbol{H}$ are given in the frame K. Because of the supposition of slow motion, the terms of the second order in regard to v/c (i.e., the dependence of mass on velocity) will be neglected; then the force acting on the charge is determined by the formula of Lorentz:

$$\boldsymbol{F} = e(\boldsymbol{E} + \mu_0 \boldsymbol{v} \times \boldsymbol{H}).$$

The full force acting on the infinitesimal element of the volume ΔV is the sum of the forces acting on all charges localized in this volume (electrons of conductivity or ions of the crystal lattice of a rock):

$$\mathbf{F} = \mathbf{E} \sum_{e_i \in \Delta V} e_i + \left(\sum_{e_i \in \Delta V} e_i \mathbf{v}_i \right) \times \mathbf{B}. \tag{9.22}$$

Here $\boldsymbol{v}_i$ is the velocity of the charge e_i with number i, and the fields $\boldsymbol{E}_i$ and $\boldsymbol{H}_i$ acting on this charge have been replaced by the mean values $\boldsymbol{E} = \langle \boldsymbol{E}_i \rangle$ and $\boldsymbol{B} = \mu_0 \langle \boldsymbol{H}_i \rangle$, respectively. This is possible for the case of *weak electrical and magnetic polarization*. This supposition may be considered to be justified sufficiently because the polarization is attenuated because of an increase in the temperature (of the order of 30 °C/km) due to the geothermic gradient. Let us note that the minimal depth of the location of typical conductive structures ($\sigma_e \geq 0.1$ См/м) in the seismic zones is of the order of 20 km or more.

The first sum in (9.22) is the full electrical charge of the element of volume, i.e., according to the definition, $\rho_e \Delta V$, the second sum is the resulting current $\boldsymbol{j}\Delta V$. But the medium is supposed to be electrically neutral; therefore, we get the *density* of the resulting EM force

$$\boldsymbol{f} = \boldsymbol{j} \times \boldsymbol{B}.$$

Now, because of the influence of the EM force, the equation of motion take the form

$$\mathrm{H}\mathbf{u} + \mathrm{A}_{TMD}(\mathbf{u}, \mathbf{w}) = \rho \mathbf{g}$$

where

$$\mathrm{A}_{TMD}(\mathbf{u}, \mathbf{w}) \equiv -\mathrm{Div}\ \widehat{\sigma}_T + \mathbf{B} \times \mathrm{rot}\mathbf{H}.$$

To take account of the influence of the EM field on the thermal one, we need to calculate the velocity of the heat production due to ohmic dissipation. The density q'_M of ohmic dissipation per second in the moving frame K' (in a small vicinity of the given point $\boldsymbol{x}'$, where the motions of all points may be neglected) is known (Tamm 1979): $q'_M = \mathbf{j}' \cdot \mathbf{E}'$. Dissipation of the energy of an electrical current does not depend on the motion of a conductor (Landau et al. 1984), i.e., the density of ohmic dissipation q_M is the same in K as in K' (the sizes of the time and spatial domains do not change by going over from K' to K because of the supposition about a *slowly moving medium*):

$$q_M = \boldsymbol{j}' \cdot \boldsymbol{E}' = \boldsymbol{j} \cdot (\boldsymbol{E} + \boldsymbol{v} \times \boldsymbol{B})$$

where transformations (9.17) were used by $\rho_e = 0$.

The sum $\boldsymbol{E} + \boldsymbol{v} \times \boldsymbol{B}$ is expressed from (9.18), and $\boldsymbol{j}$ may be replaced by rot $\boldsymbol{H}$ from Maxwell's equation without displacement currents $\boldsymbol{j} = \text{rot}\, \boldsymbol{H}$; then

$$\mathbf{E} + \mathbf{v} \times \mathbf{B} = \sigma_e^{-1}(\text{rot}\mathbf{H} + \pi_0 \text{ grad } \vartheta).$$

Therefore, the ohmic dissipation in K is

$$q_M = \frac{1}{\sigma_e}(\text{rot}\mathbf{H} + \pi_0 \text{ grad } \vartheta) \cdot \text{rot}\mathbf{H}.$$

This heat source must be added into the equation of heat conductivity, i.e., the second equation in system (9.11) with the operators $\text{P}_T\vartheta$ and $\text{B}_{DT}(\mathbf{u})$ of the left-hand side determined by (9.15) and (9.16). Then the modified equation of heat conductivity may be written in the following form with the operator coefficients

$$\text{P}_T\vartheta + \text{B}_{DMT}(\mathbf{u}, \mathbf{w}) = q \tag{9.23}$$

where

$$\text{B}_{DMT}(\mathbf{u}, \mathbf{w}) \equiv \text{B}_{DT}(\mathbf{u}) - \frac{1}{\sigma_e}(\text{rot}\mathbf{H} + \pi_0 \text{ grad } \vartheta) \cdot \text{rot}\mathbf{H}. \tag{9.24}$$

So, to describe mechanical–EM–thermal interaction by small deformations and temperature changes in a nonuniform, anisotropic, heat and electrically conductive, neutral, viscoelastic, slowly moving medium with negligible dielectric and magnetic polarization and displacement currents, we have the following S.-D. system of equations of motion, magnetic field, and heat transfer:

$$\begin{aligned}
&\mathrm{H}\mathbf{u} - \mathrm{Div}\ \widehat{\sigma}_T + \mathbf{B} \times \mathrm{rot}\mathbf{H} = \rho\mathbf{g} \\
&\mathrm{P}_M\mathbf{H} + \mathrm{rot}\left(\sigma_{\mathrm{e}}^{-1}\pi_0\ \mathrm{grad}\ \vartheta\right) - \mathrm{rot}[\mathbf{v} \times \mathbf{B}] = 0 \\
&\mathrm{P}_T(\vartheta) + \vartheta_0\frac{\partial}{\partial t}\int_0^t \phi_{ij}(t,\tau)\frac{\partial \varepsilon_{ij}}{\partial \tau}(\tau)d\tau \\
&\qquad - \frac{1}{\sigma_{\mathrm{e}}}(\mathrm{rot}\mathbf{H} + \pi_0\ \mathrm{grad}\ \vartheta)\cdot \mathrm{rot}\mathbf{H} = q.
\end{aligned} \tag{9.25}$$

Here the hyperbolic operator H is determined by (9.12) and equalities after it; the parabolic operators P_M and P_T are determined by (9.21) and (9.15), respectively.

In (9.25), the vector (three-component) equation for a magnetic field and the scalar equation for heat transfer may be combined in the form of a four-component equation

$$\mathrm{P}\boldsymbol{w} + \mathrm{B}(\boldsymbol{u}, \boldsymbol{w}) = \boldsymbol{q}$$

where $\mathrm{P} \equiv \{\mathrm{P}_M, \mathrm{P}_T\}$; operator B includes three components determined by the second and third terms of the left-hand side of the equation for magnetic field $\boldsymbol{H}$ and the fourth component is determined by the second and third terms of the left-hand side of the equation for θ; $\boldsymbol{q} \equiv (0, 0, 0, q)$. So, (9.25) is the S.-D. system (see the end of the next paragraph).

We have considered S.-D. systems where interaction of fields of different natures was described by additional terms ("sources") that had been included in equations without interaction. It is possible to consider dependence of the mechanical characteristics of a medium upon the fields of another nature localized there, i.e., dependence of the coefficients of the equation for field $\boldsymbol{u}$ upon fields of another nature $\boldsymbol{w}$, and vice versa, to include the temperature dependence in the relaxation kernels. The residual magnetization and its seismic dynamics may be considered for conductive bodies (especially a strong magnetite) in the upper floors of the crust for temperatures below the Curie point. Also, melted or partially melted conductors may be considered. The S.-D. systems for the cases of these and other suppositions (ferruginous quartzite, displacement currents) about mechanical–EM–thermal interaction are omitted here.

9.4 SDSE of Magneto-thermo-elasticity in a Nonuniform Medium

The viscoelastic models, as compared with ones without "memory"(see previous section), are more preferable from the physical viewpoint because the dissipative properties of the medium and the variants of its possible motions are represented better by these models. Indeed, because of the spatial nonuniformity of the relaxation kernels, elastic regimes and nonelastic ones with large deformations of a kind like flow (e.g., melted or partially melted metals in lower lithospheric floors) may be described by viscoelastic anisotropic models but only when the quantitative characteristics of the relaxation kernels are known. This is a difficult unsolved problem, especially for the case of tectonically active lithospheric zones (these are of most interest here) with plenty of cracks for upward motion of the melted material from the asthenosphere. In other words, because of the difficulties of the physics of rocks of seismic regions, the viscoelastic models are of interest for theoretical, physical, and mathematical investigations, but the quantitative error resulting from their application may be larger than those for more simple models with more definite input data. For these reasons and because the goal is to consider a concrete problem of the *precursory physics*—namely to calculate, in the framework of the main physical principles and geological–geophysical data, seismic generation and propagation of signals similar to ones observed before earthquakes (let us refer to these signals as *precursors*, regardless of their prognostic value, which is not known well for different seismic regions in different geodynamic states)—we now turn to the model of an *magneto-thermo-elastic, isotropic, nonuniform medium.*

So, let us suppose that *the relaxation functions of medium G, φ, and m does not depend on the past*. Then in the limited case of vanishing "memory," we go over from the system of equations of magneto-thermo-viscoelasticity (9.25) to the system of equations of MTE interaction. Equivalently, this system may be deduced from (9.11) and (9.12) if the thermo-mechanical interaction is determined according to the Hook–Duhamel–Neumann law for thermo-elastic media (Nemeth 2011; Podstrigach et al. 1984):

$$\begin{aligned} \sigma_{ij} &\equiv 2\mu\varepsilon_{ij} + \delta_{ij}\lambda \operatorname{div} \mathbf{u} \\ \mathrm{A}_{TD}(\vartheta) &\equiv \operatorname{grad}(\beta\vartheta) \end{aligned} \tag{9.26}$$

where δ_{ij} is the Kronecker delta, $\widehat{\varepsilon}$ is the tensor of deformations (9.13), and operators P_T and B_{DT} of S.-D. system (9.11) are determined by the formulas

$$\begin{aligned} \mathrm{P}_T(\vartheta) &\equiv \rho c_V \frac{\partial\vartheta}{\partial t} - \operatorname{div}(\kappa \operatorname{grad} \vartheta) \\ \mathrm{B}_{DT}(\mathbf{u}) &\equiv \beta\vartheta_0 \operatorname{div} \mathbf{v}. \end{aligned} \tag{9.27}$$

Here ρ is density; $\boldsymbol{u}$ denotes the elastic displacements; $\boldsymbol{v} = \partial\boldsymbol{u}/\partial t$; ϑ_0 is the reference temperature; μ, λ denote the elastic constants of Lame; $\beta = (2\mu + 3\lambda)\alpha$; α is the coefficient of linear thermal expansion; c_V is the heat capacity by constant volume; and κ is the coefficient of heat conductivity. The characteristics of the medium may depend on the spatial point (and on time, if $\rho c_V \partial\vartheta/\partial t$ is replaced by $\partial(\rho c_V \vartheta)/\partial t$, but we will not use this because reliable numerical data regarding nonstationary parameters of a seismic geological medium are not known).

Besides the "memory" of the medium, we have neglected anisotropy in (9.26) and (9.27). Operator $\mathrm{B}_{DT}(\boldsymbol{u})$ describes heat generation due to a change in volume by deformation (Landau and Lifshitz 1986; Podstrigach et al. 1984).

Thus, from (9.11), (9.12), (9.26), and (9.27), and repeating for consideration (9.17)–(9.23), we come to the following S.-D. system of equations for calculation of MTE interaction in an isotropic, nonuniform geological medium [see a more detailed supposition about the medium before the system of Eq. (9.25)]:

$$\begin{aligned}\mathbf{Hu} + \mathrm{A}(\mathbf{u}, \mathbf{w}) &= \rho\mathbf{g}\\ \mathbf{Pw} + \mathrm{B}(\mathbf{u}, \mathbf{w}) &= \mathbf{q}\end{aligned}$$

where the following operators are used:

$$\mathrm{A}(\mathbf{u}, \mathbf{w}) \equiv \mathrm{grad}(\beta\vartheta) + \mathbf{B} \times \mathrm{rot}\mathbf{H}$$

$$\mathbf{Hu} \equiv \rho\frac{\partial^2\mathbf{u}}{\partial t^2} - \mathrm{Div}\widehat{\sigma}$$

$$\mathbf{Pw} \equiv \begin{pmatrix} \dfrac{\partial \mathbf{B}}{\partial t} + \mathrm{rot}\left(\sigma_{\mathrm{e}}^{-1}\mathrm{rot}\mathbf{H}\right) \\ \rho c_V\dfrac{\partial \vartheta}{\partial t} - \mathrm{div}(\kappa\,\mathrm{grad}\,\vartheta) \end{pmatrix} \tag{9.28}$$

$$\mathrm{B}(\mathbf{u}, \mathbf{w}) \equiv \begin{pmatrix} -\mathrm{rot}[\mathbf{v} \times \mathbf{B}] + \mathrm{rot}\left(\sigma_{\mathrm{e}}^{-1}\pi_0\,\mathrm{grad}\,\vartheta\right) \\ \beta\vartheta_0\,\mathrm{div}\,\mathbf{v} - \sigma_{\mathrm{e}}^{-1}(\mathrm{rot}\mathbf{H} + \pi_0\,\mathrm{grad}\,\vartheta)\cdot\mathrm{rot}\mathbf{H} \end{pmatrix}.$$

Here and below, $\boldsymbol{v} = \partial \boldsymbol{u}/\partial t$ and the spatial point is the same for the characteristics of the medium and fields.

The hyperbolic operator of elasticity H is the same as for the case of the S.-D. system of thermo-elasticity, but the operator A of the influence of nonmechanical fields $\boldsymbol{w} \equiv (\boldsymbol{H}, \vartheta)$ now includes, besides the thermal term, the EM term $\boldsymbol{B} \times \mathrm{rot}\,\boldsymbol{H}$, where $\boldsymbol{B} = \mu_e\boldsymbol{H}$ is induction of the magnetic field, $\boldsymbol{H}$ is its intensity, and μ_e is magnetic permeability.

The parabolic system P $\boldsymbol{w} = \boldsymbol{q}$, describing the evolution of nonmechanical fields by $\boldsymbol{v} = 0$, includes three equations for a magnetic field (the first vector line of operator P) and a fourth equation for heat conductivity (the second scalar line of P). Operator B of the influence of nonstationary deformations on magnetic and thermal fields includes vector and scalar lines corresponding to magnetic and temperature equations of the system P$\boldsymbol{w} = \boldsymbol{q}$, contrary to the case of thermo-elasticity.

Now let us consider conditions of conjugation of the EM field components on the boundary of the two parts (with the different physical properties) of a slowly moving medium. We believe that in the frame K' moving along with the medium

$$\boldsymbol{B}' = \mu_e\,\boldsymbol{H}', \boldsymbol{D}' = \varepsilon_e\,\boldsymbol{E}'$$

where μ_e and ε_e are the magnetic and electrical permeability, respectively. Then we have from the formulas of transformation (9.17) of the fields by going over from the frame K' to the laboratory frame K:

$$\boldsymbol{B} = \boldsymbol{B}' = \mu_e(\boldsymbol{H} - \boldsymbol{v} \times \boldsymbol{D}), \boldsymbol{D} = \boldsymbol{D}' = \varepsilon_e(\boldsymbol{E} + \boldsymbol{v} \times \boldsymbol{B}).$$

Let us substitute $\boldsymbol{B}$ from the first formula in the formula for $\boldsymbol{D}$ (we remember that the motion of the medium is slow and $v/c << 1$; c is the light velocity in the medium); analogically we exclude $\boldsymbol{D}$ from the formula for $\boldsymbol{B}$:

$$\begin{aligned}
\mathbf{D} &= \varepsilon_e\mathbf{E} + \varepsilon_e\mu_e\mathbf{v} \times (\mathbf{H} - \mathbf{v} \times \mathbf{D}) \\
&= \varepsilon_e\mathbf{E} + c^{-2}\mathbf{v} \times \mathbf{H} - c^{-2}\mathbf{v} \times [\mathbf{v} \times \mathbf{D}] \\
&= \varepsilon_e\mathbf{E} + O\left((v/c)^2\right) \\
\mathbf{B} &= \mu_e\mathbf{H} - \varepsilon_e\mu_e\mathbf{v} \times (\mathbf{E} + \mathbf{v} \times \mathbf{B}) \\
&= \mu_e\mathbf{H} - c^{-2}\mathbf{v} \times \mathbf{E} - c^{-2}\mathbf{v} \times [\mathbf{v} \times \mathbf{B}] \\
&= \mu_e\mathbf{H} + O\left((v/c)^2\right).
\end{aligned} \tag{9.29}$$

Let us suppose that *surface charges and currents are absent*; then the conditions of conjugation of fields on the boundary of parts 1 and 2 of the medium in frame K', moving along with the medium, are written in the form

$$\begin{aligned}
\left(\mathbf{E}_1' - \mathbf{E}_2'\right) \times \mathbf{n} &= 0 \\
\left(\mathbf{H}_1' - \mathbf{H}_2'\right) \times \mathbf{n} &= 0 \\
\left(\mathbf{D}_1' - \mathbf{D}_2'\right) \cdot \mathbf{n} &= 0 \\
\left(\mathbf{B}_1' - \mathbf{B}_2'\right) \cdot \mathbf{n} &= 0
\end{aligned} \tag{9.30}$$

where $\boldsymbol{n}$ is normal to the boundary in frame K and indices «1» and «2» are for the field in parts 1 and 2 of the medium, respectively; $\boldsymbol{n}'$ (normal in K') has been replaced by $\boldsymbol{n}$ because the scale is the same for K and K' by slow motion.

Let us express fields in moving system K' through their values in K; see (9.17):

$$\begin{aligned}
\left(\mathbf{E}_1' - \mathbf{E}_2'\right) \times \mathbf{n} &= (\mathbf{E}_1 + \mathbf{v} \times \mathbf{B}_1 - \mathbf{E}_2 - \mathbf{v} \times \mathbf{B}_2) \times \mathbf{n} = \mathbf{0} \\
\left(\mathbf{H}_1' - \mathbf{H}_2'\right) \times \mathbf{n} &= (\mathbf{H}_1 - \mathbf{v} \times \mathbf{D}_1 - \mathbf{H}_2 + \mathbf{v} \times \mathbf{D}_2) \times \mathbf{n} = \mathbf{0}.
\end{aligned}$$

We rewrite these equalities in the form

$$\begin{aligned}
(\mathbf{E}_1 - \mathbf{E}_2) \times \mathbf{n} &= (-\mathbf{v} \times \mathbf{B}_1 + \mathbf{v} \times \mathbf{B}_2) \times \mathbf{n} \\
(\mathbf{H}_1 - \mathbf{H}_2) \times \mathbf{n} &= (\mathbf{v} \times \mathbf{D}_1 - \mathbf{v} \times \mathbf{D}_2) \times \mathbf{n}
\end{aligned}$$

and apply the vector identity

$$\boldsymbol{a} \times [\boldsymbol{b} \times \boldsymbol{c}] = \boldsymbol{b}(\boldsymbol{a} \cdot \boldsymbol{c}) - \boldsymbol{c}(\boldsymbol{a} \cdot \boldsymbol{b}).$$

Then we obtain

$$\begin{aligned}
(\mathbf{E}_1 - \mathbf{E}_2) \times \mathbf{n} &= -\mathbf{n} \times [\mathbf{v} \times (\mathbf{B}_2 - \mathbf{B}_1)] = \mathbf{n} \times [(\mathbf{B}_2 - \mathbf{B}_1) \times \mathbf{v}] \\
&= (\mathbf{n} \cdot \mathbf{v})(\mathbf{B}_2 - \mathbf{B}_1) - \mathbf{v}(\mathbf{n} \cdot (\mathbf{B}_2 - \mathbf{B}_1)) \\
&= v_n(\mathbf{B}_2 - \mathbf{B}_1), \\
(\mathbf{H}_1 - \mathbf{H}_2) \times \mathbf{n} &= -\mathbf{n} \times [\mathbf{v} \times (\mathbf{D}_1 - \mathbf{D}_2)] = \\
&= (\mathbf{n} \cdot \mathbf{v})(\mathbf{D}_1 - \mathbf{D}_2) - \mathbf{v}(\mathbf{n} \cdot (\mathbf{D}_1 - \mathbf{D}_2)) \\
&= v_n(\mathbf{D}_1 - \mathbf{D}_2)
\end{aligned}$$

because, due to as a result of (9.17) and (9.30),

$$\begin{aligned}
\mathbf{n} \cdot (\mathbf{B}_2 - \mathbf{B}_1) &= \mathbf{n} \cdot \left(\mathbf{B}_2' - \mathbf{B}_1'\right) = 0 \\
\mathbf{n} \cdot (\mathbf{D}_2 - \mathbf{D}_1) &= \mathbf{n} \cdot \left(\mathbf{D}_2' - \mathbf{D}_1'\right) = 0.
\end{aligned}$$

Here and below, $v_n \equiv (\boldsymbol{v} \cdot \boldsymbol{n})$ is the component of the velocity normal to the boundary.

Substituting (9.29) in the conditions for the tangential components of intensities, we find

$$\begin{aligned}
(\mathbf{E}_1 - \mathbf{E}_2) \times \mathbf{n} &= \mu_e v_n(\mathbf{H}_2 - \mathbf{H}_1) + O\left((v/c)^2\right) \\
(\mathbf{H}_1 - \mathbf{H}_2) \times \mathbf{n} &= \varepsilon_e v_n(\mathbf{E}_1 - \mathbf{E}_2) + O\left((v/c)^2\right).
\end{aligned} \tag{9.31}$$

These equalities give, after the vector multiplication by $\boldsymbol{n}$,

$$\begin{aligned}
\mathbf{n} \times [(\mathbf{E}_1 - \mathbf{E}_2) \times \mathbf{n}] &= -\mu_e v_n(\mathbf{H}_2 - \mathbf{H}_1) \times \mathbf{n} + O\left((v/c)^2\right) \\
&= -(v_n/c)^2(\mathbf{E}_1 - \mathbf{E}_2) + O\left((v/c)^2\right) \\
\mathbf{n} \times [(\mathbf{H}_1 - \mathbf{H}_2) \times \mathbf{n}] &= -\varepsilon_e v_n(\mathbf{E}_1 - \mathbf{E}_2) \times \mathbf{n} + O\left((v/c)^2\right) \\
&= -(v_n/c)^2(\mathbf{H}_1 - \mathbf{H}_2) + O\left((v/c)^2\right)
\end{aligned}$$

where $c = \sqrt{1/(\mu_e \varepsilon_e)}$ is the light velocity in the medium. Applying once more the identity $\boldsymbol{a} \times [\boldsymbol{b} \times \boldsymbol{c}] = \boldsymbol{b}(\boldsymbol{a} \cdot \boldsymbol{c}) - \boldsymbol{c}(\boldsymbol{a} \cdot \boldsymbol{b})$,we have, for an arbitrary vector $\boldsymbol{f}$,

$$\boldsymbol{n} \times [\boldsymbol{f} \times \boldsymbol{n}] = \boldsymbol{f} - \boldsymbol{n}(\boldsymbol{n} \cdot \boldsymbol{f}) = \boldsymbol{f}_{\parallel}.$$

Here $\boldsymbol{f}_{\parallel}$ denotes the tangential (to the boundary) component. Therefore,

$$
\begin{aligned}
(\mathbf{E}_1 - \mathbf{E}_2)_{\|} &= -(v_n/c)^2(\mathbf{E}_1 - \mathbf{E}_2) + O\left((v/c)^2\right) \\
(\mathbf{H}_1 - \mathbf{H}_2)_{\|} &= -(v_n/c)^2(\mathbf{H}_1 - \mathbf{H}_2) + O\left((v/c)^2\right).
\end{aligned}
$$

Multiplying once more by $\boldsymbol{n}$ and taking into account that $\boldsymbol{f}_{\|} \times \boldsymbol{n} = \boldsymbol{f} \times \boldsymbol{n}$, we come to

$$
\begin{aligned}
(\mathbf{E}_1 - \mathbf{E}_2) \times \mathbf{n} &= O\left((v/c)^2\right) \\
(\mathbf{H}_1 - \mathbf{H}_2) \times \mathbf{n} &= O\left((v/c)^2\right).
\end{aligned}
$$

So, because of the supposition about *slow motion and negligible surface charges and currents*, we shall consider, on contacts of parts with different physical properties of a moving medium, the usual conditions of the field conjugation:

$$
\begin{aligned}
(\mathbf{E}_1 - \mathbf{E}_2) \times \mathbf{n} &= 0 \\
(\mathbf{H}_1 - \mathbf{H}_2) \times \mathbf{n} &= 0 \\
(\mathbf{D}_1 - \mathbf{D}_2) \cdot \mathbf{n} &= 0 \\
(\mathbf{B}_1 - \mathbf{B}_2) \cdot \mathbf{n} &= 0.
\end{aligned}
\tag{9.32}
$$

Let us note that because of the supposition about slow motion (namely because of its sequences $\boldsymbol{B} = \mu_e \boldsymbol{H}$, (9.29), and div $\boldsymbol{B} = 0$) it is possible the vector line of operator P$\boldsymbol{w}$, (9.28) [or, and it is the same, the operator $\mathrm{P}_M(\boldsymbol{w})$ of magnetic diffusion (9.21)], to write, in the traditional form for parabolic operators,

$$
\begin{aligned}
\mathrm{P}_M(\mathbf{w}) &\equiv \frac{\partial \mathbf{B}}{\partial t} + \mathrm{rot}\left(\sigma_e^{-1} \mathrm{rot}\mathbf{H}\right) \\
&= \mu_e \frac{\partial \mathbf{H}}{\partial t} + \sigma_e^{-1} \mathrm{rot}\,\mathrm{rot}\mathbf{H} + \left(\mathrm{grad}\, \sigma_e^{-1}\right) \times \mathrm{rot}\mathbf{H} \\
&= \mu_e \frac{\partial \mathbf{H}}{\partial t} - \sigma_e^{-1} \Delta \mathbf{H} + \sigma_e^{-1} \mathrm{grad}\, \mathrm{div}\, \mathbf{H} + \left(\mathrm{grad}\, \sigma_e^{-1}\right) \times \mathrm{rot}\mathbf{H} \\
&= \mu_e \frac{\partial \mathbf{H}}{\partial t} - \sigma_e^{-1} \Delta \mathbf{H} + \sigma_e^{-1} \mathrm{grad}\, \mathrm{div}\left(\mu_e^{-1} \mathbf{B}\right) + \left(\mathrm{grad}\, \sigma_e^{-1}\right) \times \mathrm{rot}\mathbf{H} \\
&= \mu_e \frac{\partial \mathbf{H}}{\partial t} - \sigma_e^{-1} \Delta \mathbf{H} + \sigma_e^{-1} \mathrm{grad}\left(\mathbf{B}\ \mathrm{grad}\, \mu_e^{-1}\right) + \left(\mathrm{grad}\, \sigma_e^{-1}\right) \times \mathrm{rot}\mathbf{H}
\end{aligned}
$$

where we have used div $\boldsymbol{B} = 0$.

Evidently, terms with $\partial \boldsymbol{H}/\partial t$ and $\Delta \boldsymbol{H}$ determine the parabolic type (Ladyjenskaya et al. 1967) of the operator of magnetic diffusion, whereas the last two terms are operators of the first order ($\boldsymbol{B} = \mu_e \boldsymbol{H}$ because of slow motion, let us remember) in regard to the magnetic field.

Thus, the supposition about slow motion provides the S.-D. structure of the systems of equations of MTE interaction (9.25) and (9.28).

9.5 The 2D System of Equations of the Low-Frequency Magneto-Thermo-Elastic Interaction in the Case of a Medium Uniform in Regard of Magnetic Permeability

We shall consider a 2D case, i.e., where the fields and physical characteristics of a medium do not depend on the third coordinate z; moreover, the z-components of elastic displacements and the magnetic field are zero:

$$\begin{aligned} \gamma &= \gamma(x, y) \\ \vartheta &= \vartheta(t, x, y) \\ \mathbf{u} &= (u_1(t, x, y), u_2(t, x, y), 0) \\ \mathbf{H} &= (H_1(t, x, y), H_2(t, x, y), 0). \end{aligned} \tag{9.33}$$

Here γ is any of the physical parameters of the medium: ρ, λ, μ, β, σ_{e}, μ_{e}, c_V, κ; see Sects. 9.3 and 9.4. We shall neglect the thermo-electric current, i.e., we shall suppose:

$$\pi_0 = 0. \tag{9.34}$$

The axis Ox is directed vertically downward; the axis Oy belongs to the horizontal plane and is directed perpendicularly to the structures, i.e., perpendicularly to the direction where the size of the structures is maximal; Oz is directed horizontally along the structures.

By this geometry,

$$\begin{aligned} \mathrm{rot}\mathbf{H} &= (0,\ \ 0,\ \ H_{2,1} - H_{1,2}), \\ \mathbf{B} \times \quad \mathrm{rot}\mathbf{H} &= ((H_{2,1} - H_{1,2})B_2,\ \ -(H_{2,1} - H_{1,2})B_1,\ \ 0). \end{aligned} \tag{9.35}$$

Therefore, the vector of the electrical current density has the only nonzero component:

$$j_{\mathrm{e}} = \{0, 0, j_{\mathrm{e}}\}, j_{\mathrm{e}} = H_{2,1} - H_{1,2}.$$

Now equations of motion (9.26) and (9.28) take the form

$$\begin{aligned} \rho u_{1,tt} - \left(\mu\left(u_{1,j} + u_{j,1}\right)\right)_{,j} - (\lambda u_{k,k})_{,1} + (\beta\vartheta)_{,1} + B_2(H_{2,1} - H_{1,2}) &= \rho g \\ \rho u_{2,tt} - \left(\mu\left(u_{2,j} + u_{j,2}\right)\right)_{,j} - (\lambda u_{k,k})_{,2} + (\beta\vartheta)_{,2} - B_1(H_{2,1} - H_{1,2}) &= 0. \end{aligned} \tag{9.36}$$

The repeated indices denote here summation from 1 to 2; the notations for derivatives were used:

$$f,_1 \equiv \partial f/\partial x, f,_2 \equiv \partial f/\partial y, f,_{tt} \equiv \partial^2 f/\partial t^2.$$

Taking account of (9.34), Eq. (9.28) may be rewritten as

$$\mathbf{B}_{,t} + \mathrm{rot}\left(\sigma_e^{-1}\mathrm{rot}\mathbf{H}\right) - \mathrm{rot}(\mathbf{v}\times\mathrm{rot}\mathbf{B}) = 0, \quad v = \partial u/\partial t.$$

Because of the 2D geometry considered now,

$$\mathrm{rot}\left(\sigma_e^{-1}\mathrm{rot}\mathbf{H}\right) = \left\{(1/\sigma_e(H_{2,1} - H_{1,2}))_{,2}, \quad - \quad (1/\sigma_e(H_{2,1} - H_{1,2}))_{,1}, \quad 0\right\}$$
$$\mathbf{v}\times\mathbf{B} = (0, \ 0, \ v_1B_2 - v_2B_1)$$
$$\mathrm{rot}[\mathbf{v}\times\mathbf{B}] = \left\{(v_1B_2 - v_2B_1)_{,2}, \quad -(v_1B_2 - v_2B_1)_{,1}, \quad 0\right\}.$$

As a result, the equations of the magnetic field dynamics may be written as

$$\begin{aligned} B_{1,t} + \left(\sigma_e^{-1}\cdot(H_{2,1} - H_{1,2})\right)_{,2} - (v_1B_2 - v_2B_1)_{,2} = 0 \\ B_{2,t} - \left(\sigma_e^{-1}\cdot(H_{2,1} - H_{1,2})\right)_{,1} + (v_1B_2 - v_2B_1)_{,1} = 0. \end{aligned} \tag{9.37}$$

Finally, because of $\pi_0 = 0$ [see (9.34)] and formulas (9.35), the equation of heat conductivity from the previous section may be simplified:

$$\rho c_V \vartheta_{,t} - (\kappa\vartheta_{,i})_{,i} + \beta\vartheta_0 v_{k,k} - \sigma_e^{-1}\cdot(H_{2,1} - H_{1,2})^2 = q. \tag{9.38}$$

Equations (9.33)–(9.38) will be used to describe MTE interaction of the fields in a nonuniform 2D lithospheric domain (the suppositions about the model lithosphere and processes considered there were formulated in two previous sections).

Below we shall neglect magnetic polarization and suppose that polarization μ_e of the medium under consideration is equal to the magnetic polarization of vacuum μ_0 everywhere in the medium:

$$\mu_e = \mu_0. \tag{9.39}$$

Indeed, the magnetic structures of dipped (typical depths are of the order of 20 km or more) low-resistivity inclusions are distructed, as a whole or partially, by high temperature: the geothermal gradient may be of the order of 30 °C/km for a seismic lithospheric zone (elevated values of this gradient are typical there) and the temperature at the depths typical for considerable conductive inclusions is near the Curie point.

Because the magnetic permeability of the medium is supposed to be uniform:

$$\mathrm{div}\boldsymbol{B} = \mathrm{div}\boldsymbol{H} = 0, \ -\partial H_1/\partial x = \partial H_2/\partial y.$$

Thus, $-H_1(t, x, y)$ and $H_2(t, x, y)$ may be regarded as partial derivatives of the "scalar magnetic potential" $A(t, x, y)$:

$$H_1 = -\partial A/\partial y, H_2 = \partial A/\partial x. \tag{9.40}$$

Substituting these expressions in Eq. (9.37) and taking into account that:

$$H_{2,1} - H_{1,2} = \partial^2 A/\partial x^2 + \partial^2 A/\partial y^2 = \Delta A,$$

we obtain two equations with the same left sides:

$$\mu_e A_{,t} - \sigma_e^{-1}\Delta A + \mu_e(v_1 A_{,1} + v_2 A_{,2}) = f_1(t, x)$$
$$\mu_e A_{,t} - \sigma_e^{-1}\Delta A + \mu_e(v_1 A_{,1} + v_2 A_{,2}) = f_2(t, y)$$

where $f_1(t, x)$ and $f_2(t, y)$ are some functions. It follows from coincidence that $f_1(t, x) = f_2(t, y)$ and the right side of the last two equations may be a function of t only:

$$\mu_e A_{,t} - \sigma_e^{-1}\Delta A + \mu_e \cdot (v_1 A_{,1} + v_2 A_{,2}) = \psi(t). \tag{9.41}$$

Function $\psi(t)$ may be chosen identically equal to zero because the magnetic field determined by "potential" A with the help of (9.40) is independent from the choice of $\psi(t)$. Indeed, the term-by-term differentiation of Eq. (9.41) by y for any $\psi(t)$ and going over from A to $\boldsymbol{H}$ according to (9.40) gives Eq. $(9.37)_1$, whereas differentiation by x gives Eq. $(9.37)_2$. Thus, the equations for the components of the magnetic field determined by (9.40) from A do not depend on the choice of $\psi(t)$.

The same is true for the initial boundary value conditions of the magnetic field; therefore, it is true for the magnetic field itself (we consider the well-posed initial boundary value problems only and the solution is unique in these conditions).

As the result, the equation for the magnetic potential disturbed by elastic displacements of the medium will be written in the form

$$\mu_e \tilde{A}_{,t} - \sigma_e^{-1}\Delta\left(A_0 + \tilde{A}\right) + \mu_e \cdot \left(v_1 H_{20} - v_2 H_{10} + v_1 \tilde{A}_{,1} + v_2 \tilde{A}_{,2}\right) = 0 \tag{9.42}$$

where $\tilde{A} = A - A_0$ denotes the deviation of the magnetic potential from its initial value A_0 by $t = 0$; H_{20} and H_{10} are the horizontal and vertical components of the undisturbed magnetic field.

Below we are interested in seismic disturbances of the magnetic field only and its constant component will not be of interest to us; the same is true for the magnetic potential seismic disturbances and A_0 will not be of interest. For these reasons, we shall write simply A instead of $\tilde{A} = A - A_0$, i.e., just beginning from this place, A is meant as the deviation of the magnetic potential from the stationary value A_0.

So in the 2D case and because of (9.39)–(9.42), the equations of motion (9.36), the magnetic field (9.37), and heat conductivity (9.38) are written in the form of the S.-D. system of PDEs with the interaction operators of the second order:

$$
\begin{aligned}
&\rho u_{1,tt} - \left(\mu\left(u_{1,j} + u_{j,1}\right)\right)_{,j} - \left(\lambda u_{k,k}\right)_{,1} + (\beta\vartheta)_{,1} + \mu_{\mathrm{e}} A_{,1}\,\Delta A = \rho g \\
&\rho u_{2,tt} - \left(\mu\left(u_{2,j} + u_{j,2}\right)\right)_{,j} - \left(\lambda u_{k,k}\right)_{,2} + (\beta\vartheta)_{,2} + \mu_{\mathrm{e}} A_{,2}\,\Delta A = 0 \\
&\mu_{\mathrm{e}} A_{,t} - \sigma_{\mathrm{e}}^{-1}\Delta A + \mu_{\mathrm{e}}\left(v_1 A_{,1} + v_2 A_{,2}\right) = \mu_{\mathrm{e}}\left(v_2 H_{10} - v_1 H_{20}\right) \\
&\rho c_V \vartheta_{,t} - \left(\kappa\vartheta_{,i}\right)_{,i} + \beta\vartheta_0 v_{k,k} - \sigma_{\mathrm{e}}^{-1}\cdot(\Delta A)^2 = q
\end{aligned}
\tag{9.43}
$$

where $\boldsymbol{v} = \partial \boldsymbol{u}/\partial t$; summation over repeated indices is meant from 1 to 2.

References

Alfven, H. (1981). *Cosmic plasma, Astrophysics and space science library* (p. 122). New York: Springer.

Christensen, R. (1982). *Theory of viscoelasticity*. Amsterdam: Elsevier.

Feynman, R. P., Leighton, R. B., & Sands, M. (2005). *Feynman lectures on physics*. Boston: Addison–Wesley. http://www.feynmanlectures.caltech.edu. Accessed 08 Jan 2019.

Hutter, K., & Ven, A. A. F. (1978). *Field matter interactions in thermoelastic solids, Lecture notes in physics* (Vol. 88, p. 230). New York: Springer.

Iesan, D., & Quintanilla, R. (2013). On strain gradient theory of thermoviscoelasticity. *Mechanics Research Communication, 48*, 52–58.

Kashinzev, G. (2001). Geodynamics and magmatism of the initial stage of formation of Atlantic. *Геотектоника (Geotectonics), 2*, 64–77 (In Russian).

Keylis-Borok, V. I., & Monin A. S. (1959). *Magnetoeleastic waves and the boundary of the earth core*. In Proceedings of Academy of Sciences of USSR (pp. 1530–1541). Ser. Geophysics. No 11. (In Russian. Кейлис-Борок В.И., Монин А.С. 1959. Магнитоупругие волны и граница земного ядра. Известия АН СССР. Сер. Геофизическая. 1959. № 11. С. 1530–1541).

Knopoff, L. (1955). The interaction between elastic wave motion and a magnetic field in electrical conductors. *Journal of Geophysical Research, 60*(4), 441–446.

Ladyjenskaya, O. A. (1969). *The mathematical theory of viscous incompressible flow*. New York: Gordon and Breach, Science Publications. (Translated from Russian, Moscow: Gos. Izdat. Fiz-Mat Lit. 1961).

Ladyjenskaya, O. A., Solonnikov, V. A., & Uraltseva, N. N. (1967). *Linear and quasilinear equations of parabolic type. Translations of mathematical monographs* (Vol. 23). Providence: American Mathematical Society. (Translated from Russian, Moscow: Izdat. Nauka 1964).

Lakes, R. S. (2009). *Viscoelastic materials*. Cambridge: Cambridge University Press.

Landau, L. D., & Lifshitz, E. M. (1975). *The classical theory of fields* (Vol. 2, 4th ed.). Oxford: Butterworth–Heinemann.

Landau, L. D., & Lifshitz, E. M. (1980). *Statistical physics, part 1* (Vol. 5, 3rd ed.). Oxford: Butterworth–Heinemann.

Landau, L. D., & Lifshitz, E. M. (1986). *Theory of elasticity* (Vol. 7, 3rd ed.). Oxford: Butterworth–Heinemann.

Landau, L. D., & Lifshitz, E. M. (1987). *Fluid mechanics* (Vol. 6, 2nd ed.). Oxford, Butterworth–Heinemann.

Landau, L. D., Lifshitz, E. M., & Pitaevskii, L. P. (1984). *Electrodynamics of continuous media* (Vol. 8, 1st ed.). Oxford: Butterworth–Heinemann.

Lokshin, A. A., & Suvorova, Yu. V. (1982). *Mathematical theory of propagation of waves in media with memory* (p. 150). Moscow: Publishing House of Moscow State University. (In Russian. Локшин А.А., Суворова Ю.В. Математическая теория распространения волн в средах с памятью. М.: Изд-во МГУ, 1982).

Maugin, G. (1988). *Continuum mechanics of electromagnetic solids*. North Holland: Elsevier.
Mikhaylov, I. G., Solov'ev, V. A., & Syrnikov, Y. P. (1964). *Osnovy molecularnoi acustiki (Foundations of molecular acoustics)*. Moscow: Nauka. (In Russian).
Nemeth, M. D. (2011). *An in-depth tutorial on constitutive equations for anisotropic materials* (p. 582). NASA/TM-2011-217314. http://www.sti.nasa.gov. Accessed 08 Jan 2019.
Novik, O. B. (1969). The Cauchy problem for the system of PDEs including the hyperbolic and parabolic operators. *Journal of Computational Mathematics and Mathematical Physics, 1*, 122–136. (In Russian).
Novik, O. B. (1983a). *Interaction of the geothermic field with the field of deformations of a layer*. Geothermic investigations in Middle Asia and Kasahstan. Abstracts of reports of all-union conference. Ashgabat (In Russian).
Novik, O. B. (1983b). *Mathematical description of non-stationary fields of deformations and temperature*. In Proceedings of high educational institutes (Известия вузов) (p. 24). Ser. Geology and Prospecting. Deposited in VINITI, № 6788-83 (In Russian).
Novik, O. B. (1988). *Finite difference schemes for the problems of thermo-mechanical effect in the oceanic lithosphere: Collection of scientific papers* (pp. 89–97). Geothermic investigations on the bottoms of the water areas. Moscow: Nauka (Наука) (In Russian).
Novik, O. B., Mikhaylovskaya, I. B., & Ershov, S. V. (2000). *Mathematical problems of designing the development of marine geothermal deposits. Collected volume "Earth's thermal field and methods of its investigation"* (pp. 306–310). Moscow: Publishing House of Peoples' Friendship University of Russia. (In Russian: Новик О.Б., Михайловская И.Б., Ершов С.В. 2000. Математические задачи при проектировании разработки морских геотермальных месторождений. Сб. «Тепловое поле Земли и методы его изучения», стр. 306-310. Изд-во Рос. Универ. Дружбы Народов. Москва).
Podstrigach, Y. C., Lomakin, V. A., & Kolyano, Y. M. (1984). *Thermo-elasticity of bodies of non-uniform structure* (p. 368). Moscow: Nauka. (In Russian).
Shashkov, A.G., Bubnov, V.A., & Yanovsky, S.Y. (2004). *Wave phenomena of heat conductivity: the system-structural approach* (p. 289). Moscow: Editorial URSS. (In Russian).
Shirong, M. (1992). Progress in earthquake prediction in China during 80's. *Journal of Earthquake Prediction Research, 1*, 43–57.
Tamm, I. E. (1979). Fundamentals of the theory of electricity. Moscow: Mir. (Translated from Russian by G. Leib).

Chapter 10
Numerical Stability of the Discretized Initial Boundary Value Problems of the String–Diffusion Systems of Equations (SDSE)

Abstract The string–diffusion systems of equations (SDSE), being formulated correctly from a physical viewpoint (see Chap. 9), are not partial differential equations (PDEs; scalar or vector ones) for which the type may be determined in the frame of the classical typing of PDEs of mathematical physics, because this typing was formed when fields of different natures were investigated mostly without taking account of their interaction. But precursors to earthquakes (one of the main subjects of this book) are generated because of interactions of physical fields of different natures. Therefore, the finite difference and Galerkin methods for initial boundary value problems of SDSE are considered in this chapter.

Keywords Interaction of physical fields of different natures · Stability of an algorithm

10.1 Absence of a Well-Investigated Type and Nonusual Properties of SDSE

We move on to the construction and mathematical investigation of algorithms of computations of interaction of physical fields of different natures disturbed (activated) during an earthquake formation period. These algorithms are determined by the discrete analogs of the initial boundary value problems of string–diffusion systems of equations (SDSE; see Chap. 9). As the discrete analogs, the finite difference and Galerkin approximations of these problems are introduced. Then we show, with the known method of "energy inequalities" (i.e., mostly with integration by parts and integral inequalities), that some norms of the solutions of the discrete problems are restricted by the values determined by the problems' input data. As a matter of fact, these values are the same for arbitrarily small grid steps by the finite difference method or an arbitrary large number of elements of the Galerkin basis. In other words, the numerical stability of the solution has been proved for both types of discrete approximations of the initial boundary value problems under consideration. This is sufficient for practical applications of these discrete solutions,

O. Novik et al., *Electromagnetic Geophysical Fields*,
https://doi.org/10.1007/978-3-319-98461-2_10

and we intend to omit here other interesting mathematical results connected with the mentioned ones and, therefore, to omit the methods of functional analysis.

It has been shown that evolution of the interaction of physical fields of different natures in the lithosphere is described by systems of partial differential equations (PDEs) including hyperbolic and parabolic subsystems coupled by the operators of interaction of the fields. We have called these systems of PDEs string–diffusion systems of equations (SDSE; see the definition given in Chap. 9). In the parts of a medium where the interaction is attenuated, the SDSE is represented by two uncoupled subsystems: one of a hyperbolic type and another of a parabolic type. Therefore, the SDSE are not of a hyperbolic or parabolic type, though both types are partial cases of the SDSE type. The SDSE are not systems of the Cauchy–Kovalevskaya type, i.e., they are not of the form

$$D_t^m u = \sum_{\substack{|\alpha| + j \leq m \\ j \leq m-1}} a_{\alpha j}(t,x) D_x^\alpha D_0^j u + f(t,x).$$

Here $u = (u_1, \ldots, u_N), f = (f_1, \ldots, f_N)$; $a_{\alpha j}$ denotes $N \times N$ matrices; and $\alpha = (\alpha_1, \ldots, \alpha_n)$ is a vector with non-negative integer components.

$$|\alpha| = \sum_{j=1}^{n} \alpha_j,$$

derivatives are written as follows:

$$D_x^\alpha \equiv D_{x_1}^{\alpha_1} D_{x_2}^{\alpha_2} \cdots D_{x_n}^{\alpha_n},$$
$$D_{x_k}^p \equiv \left(\frac{1}{i}\frac{\partial}{\partial x_k}\right)^p,$$
$$D_t^p \equiv D_0^p \equiv \left(\frac{1}{i}\frac{\partial}{\partial t}\right)^p.$$

The parabolic equations, e.g., the heat conductivity equation $u/\partial t = \partial^2 u/\partial x^2$, belong to the class (type) of the SDSE but do not belong to the class of operators of the main type (Egorov 1984), i.e., to the class of operators

$$P(D), D = (D_0, D_1, D_2, D_n),$$

for which the symbol of the uniform main part $P_m(\xi)$ has a nonzero gradient in regard to $\xi = (\xi_0, \xi_1, \ldots, \xi_n)$ everywhere in R^{n+1} except for the point $\xi = 0$ [m is the differential order of the operator $P(D)$].

Let us note that the simplest SDSE of the theory of thermo-elasticity (see Chap. 9) may be reduced to one equation, but its operator does not belong to the main type. Indeed, let us consider the SDSE for the one-dimensional (1D) case $x = x_1$ and constant coefficients:

$$\mathrm{H}u + \mathrm{A}w = 0,$$
$$\mathrm{P}w + \mathrm{B}u = 0.$$

The operators commute in this case:

$$\mathrm{PH}u + \mathrm{PA}w = 0$$
$$\mathrm{AP}w + \mathrm{AB}u = 0.$$

Therefore,

$$(\mathrm{PH} - \mathrm{AB})u = 0.$$

Now let us consider the operator coefficient of the SDSE of thermo-elasticity:

$$\mathrm{H} = \partial_1^2 - c_1^{-2}\,\partial_t^2 = -D_1^2 + c_1^{-2}D_0^2,$$
$$\mathrm{A}w = m\partial_1 w = m\ \ iD_1\ \ w,$$
$$\mathrm{P}w = \left(-\partial_1^2 + \frac{1}{\kappa}\partial_t\right)\ w = \left(D_1^2 + \frac{i}{\kappa}D_0\right)w,$$
$$\mathrm{B}u = \eta\partial_1\partial_t u = -\eta D_1 D_0 u,$$

where

$$D_1 = -i\partial_1 = -i\partial_{x_1},\quad D_0 = D_t = -i\partial_t.$$

Then

$$(\mathrm{PH} - \mathrm{AB})u = \left\{\left[D_1^2 + (i/\kappa)D_0\right]\left[-D_1^2 + c_1^{-2}D_0^2\right] - i\eta m D_1^2\ D_0\right\}u = 0.$$

The last equation is the consequence of the thermo-elasticitySDSE under consideration; we suppose that all functions here are determined everywhere in R^1 for $t > 0$, and w may be determined, for example, by the convolution of the Green function of the operator of thermal conductivity P with the "source"—Bu.

Let us write out the main part of the symbol of the last equation of the fourth order with regard to the function $u = u(t, x_1)$:

$$\mathrm{P}_4(D) = -D_1^4 + c_1^{-2}D_1^2D_0^2$$
$$\mathrm{P}_4(\xi) = -\xi_1^4 + c_1^{-2}\,\xi_1^2\xi_0^2.$$

Then we obtain (e_1 and e_0 are the orts in R^2)

$$\mathrm{gradP}_4(\xi) = \left(-4\xi_1^3 + 2c_1^{-2}\xi_1\xi_0^2\right)e_1 + 2c_1^{-2}\xi_1^2\xi_0 e_0.$$

The operator (PH − AB) is the operator of the main type if and only if

$$| \operatorname{grad} P_4(\xi) | \neq 0, \xi \in R^2/0.$$

Evidently, this condition is not true for the equation (PH − AB)$u = 0$ of the string (conservative) component of the solution $[u(t, x_1), w(t, x_1)]$.

Because of the absence of a classical type, the SDSE have a particular set of properties that are impossible for parabolic or hyperbolic systems of equations. For example, the SDSE have an unlimited velocity of propagation of disturbances both for the diffusion component w of solution (u, w) (w is described by the parabolic operator P in the case of absence of interaction) and for the string component$u$ (described by the hyperbolic operator H). Indeed, a local disturbance of the string component u changes the term Bu of the equation Pw + Bu = 0 of the SDSE; this change causes, in turn, a change of w, and this change is nonlocal because $w = -\mathrm{P}^{-1}$ Bu is determined by the convolution of Bu with the Green function of the equation of heat transfer. Then the change of the term Aw of the equation Hu + Aw = 0 is nonlocal as well and causes a nonlocal change of u.

At the same time, the theorems of the Liouville type (the simplest example: a bounded entire function is constant) are not valid for the SDSE, because the hyperbolic equations belong to the SDSE type. Besides, the discontinuities of the initial data may disappear after some time because the parabolic equations belong to the SDSE, and may be preserved as well because this is true for the hyperbolic SDSE equations. These and other important problems of the theory of waves in dissipative media (wave fronts, Huygens' principle, superposition of the resolving operator of Fourier of a hyperbolic equation with other operator coefficients of the SDSE, etc.) are not considered here (Babich 2009; Egorov 1984).

Because of the nonusual properties (as compared with equations of classical types) of the SDSE, it is necessary to investigate the discrete approximations of the initial boundary value problems of these equations to investigate seismic disturbances of the electromagnetic (EM) and thermal fields.

The numerical result for the SDSE with constant coefficients, one spatial coordinate, and the operators H, A, and B of the first order was described in Rihtmayer and Morton (1994), but without investigation of the stability of the methods applied.

The correct solvability of the Cauchy problem of the SDSE with the operators H, A, and B of the first order and coefficients depending on $x = (x_1, x_2, \ldots x_n)$ was proved in the class of functions of the maximal order of growth by $|x| \to \infty$, compatible with the uniqueness of the solution, in Novik (1969). To get this result, numerical stability and convergence of the explicit finite difference approximation of the Cauchy problem were demonstrated. The method of the finite difference energy estimations (Ladyjenskaya et al. 1967; Ladyjenskaya 1969) in the grid-weighted Sobolev norms with parameters was used (e.g., because of the choice of the parameter of the weight, the norms are definite for functions growing by $|x| \to \infty$ as $\exp\{ax^2\}$, $a > 0$).

With the limit transition from explicit grid approximation, the correct solvability of the Cauchy problem in the class of growing functions was proved for the SDSE with hyperbolic and parabolic operators of the second order, by Mikhaylovskaya and Novik (1979). The stability and convergence of the explicit grid approximation in the case of the initial boundary value problem was investigated by Novik (1983a, b) and by Mikhaylovskaya and Novik (1995). The stability of the grid scheme when the hyperbolic operator of the SDSE was approximated explicitly and the parabolic operator was approximated by the method of fractional steps was demonstrated by Novik (1988). The Galerkin method for the two-dimensional (2D) and three-dimensional (3D) initial boundary value problems of the nonlinear dynamical theory of magneto-thermo-elasticity was investigated by Novik (1993, 1994a, b, 1995). These results are discussed below and, on their basis, the computations of the seismic disturbances of EM and thermal fields in model geological media with low-resistivity structures are fulfilled.

10.2 Semi-implicit Finite Difference Approximation

10.2.1 Initial Boundary Value Problem

It was noted in Chap. 9 that nonstationary interaction of physical fields of different natures in solids and fluids is described by systems of equations of the form

$$\mathrm{H}u + \mathrm{A}(u, v) = f \tag{10.1}$$

$$\mathrm{P}v + \mathrm{B}(u, v) = \phi. \tag{10.2}$$

Similar systems were called string–diffusion systems of equations (SDSE) by us; for a more precise definition, see Chap. 9.

In this section, linear systems are considered.

$$\mathrm{H}u + \mathrm{A}v = f, \tag{10.3}$$

$$\mathrm{P}v + \mathrm{B}u = \phi. \tag{10.4}$$

Let us remember that equations $\mathrm{H}u = f$ and $\mathrm{P}v = \phi$ of the hyperbolic (H) and parabolic (P) types, respectively, describe processes of conservative (string, S) and dissipative (diffusion, D) characters, respectively by neglect of their interaction, the operators A and B describe linear string–diffusion interaction here.

Below, the discretization of the initial boundary value problem is considered:

$$\begin{aligned}\ddot{u} &= H_{ij}(t)u_{x_ix_j} - A_{ij}(t)v_{x_i}\\&\quad - \int_0^t \left[\mathrm{H}_{ij}(t,t')u_{x_ix_j}(t') + \mathrm{A}_{ij}(t,t')v_{x_ix_j}(t')\right]\mathrm{d}t' + f\end{aligned} \tag{10.5}$$

$$\begin{aligned}\dot{v} &= P_{ij}(t)v_{x_ix_j} - B_{0i}(t)\dot{u}_{x_i} - B_{ij}(t)u_{x_ix_j}\\&\quad - \int_0^t \left(\mathrm{P}_{ij}(t,t')v_{x_ix_j}(t') + \mathrm{B}_{ij}(t')u_{x_ix_j}(t')\right)\mathrm{d}t' + \varphi\end{aligned} \tag{10.6}$$

$$\begin{aligned}&u(t,x)|_{t=0} = u_0(x), \quad x\in\Omega\\&\dot{u}(t,x)|_{t=0} = u_1(x), \quad x\in\Omega\\&v(t,x)|_{t=0} = v_0(x), \quad x\in\Omega\\&u(t,x)|_{\partial\Omega} = 0, \quad 0\le t\le T\\&v(t,x)|_{\partial\Omega} = 0, \quad 0\le t\le T.\end{aligned} \tag{10.7}$$

Here Ω is a domain in R^n, mes $\Omega < \infty$, $x = (x_1, x_2, ..., x_n)$; $u(t, x)$ and $v(t, x)$ are the real unknown vector fields with m and m' components, respectively, which must be determined in Ω from Eqs. (10.5) and (10.6), and initial boundary value conditions (10.7); $\partial\Omega$ is the notation for the boundary of Ω; m, m', and n are the arbitrary natural numbers; the coefficients of the system of Eqs. (10.5) and (10.6) are the matrices with suitable dimensions for u, v; these matrices depend on (t, x) in the case of matrices outside an integral and the matrices depend on (t, t', x) if they are under the sign of an integral; dependence of matrices on x is possible though it is not written out. Summation over repeated indices from 1 to n is meant if the opposite supposition has not been formulated.

Different linear systems of equations of the nonstationary interaction of elastic, thermal, and EM fields can be written in this form (up to lower terms) for anisotropic, nonuniform media with smooth dependence of the physical characteristics on the coordinates (see Chap. 9).

10.2.2 Grid Approximation

In this section, the following notations are used for the grid domains and steps:

R_h^n—n-dimensional cubic lattice with the step h.
$\overline{\Omega}$—closure of set Ω.
$\overline{\Omega}_h$—union (U) of all closed elementary (minimal) cubes of R_h^n belonging to $\overline{\Omega}$.
Ω_h—set of the nodes of R_h^n, which are the internal points of $\overline{\Omega}_h$.
$\widetilde{\Omega}_h$—set of nodes of R_h^n belonging to $\overline{\Omega}_h$.
τ—step in time; $\delta = \tau/n$; $T = N\tau$, N is a natural number.

The arguments of the grid functions and operators of the shifts of the time and spatial grid arguments are denoted as follows:

$$
\begin{gathered}
w \equiv w(s) = w(s\tau, x) \\
w^{\pm} \equiv w((s \pm 1)\tau, x) \\
x^{\pm i} \equiv \left(x^{(1)}, \ldots, x^{(i-1)}, x^{(i)} \pm h, x^{(i+1)}, \ldots, x^{(n)}\right) \\
w^{\pm i} \equiv w(s\tau, x^{\pm i}) \\
w(s_i) \equiv w(s\tau + i\delta, x), \qquad i = 1,\ 2,\ \ldots,\ n \\
\tilde{w}(s_i) \equiv \frac{w(s_i) + w(s_{i-1})}{2} \\
W(s, s') \equiv W(s\tau, s'\tau, x).
\end{gathered}
$$

The finite difference spatial ratios (i.e., in regard to spatial arguments) are denoted as follows:

$$
\begin{gathered}
w_{;i} \equiv \frac{w^{+i} - w}{h} \\
w_{;\bar{i}} \equiv \frac{w - w^{-i}}{h} \\
w_{;ij}^{-i} \equiv w_{;\bar{i}\bar{j}} \\
\left|w_{;x}\right|^2 \equiv \sum_i \left(\left|w_{;i}\right|^2 + \left|w_{;\bar{i}}\right|^2\right) \\
w_{;\mathring{i}} \equiv \frac{w_{;i} + w_{;\bar{i}}}{2}.
\end{gathered}
$$

The finite difference time ratios (i.e., in regard to the temporary argument) are denoted analogically:

$$
\begin{gathered}
w_{;t} = \frac{w^{+} - w}{\tau} \\
w_{;\bar{t}} = w_{;t}^{-} \\
w_{;t\bar{t}} = \frac{w_{;t} - w_{;\bar{t}}}{\tau}.
\end{gathered}
$$

Here and below, a semicolon before the lower index denotes the discrete "derivative" (i.e., the finite difference ratio) in regard to the coordinate relevant to this index; the discrete "derivative" in regard to time is denoted with a semicolon before the index t.

Contrary to the operators of the shifts, the involution (positive powers are meant only) is denoted by the upper index without a sign.

Let us approximate (10.5) explicitly and (10.6), with the help of fractional steps; then the following system of grid equations will be obtained:

$$
\begin{aligned}
u_{;t\bar{t}}(s) - H_{ij}(s)u_{;\bar{i}\bar{j}}(s) + A_i(s)\tilde{v}_{;i}\big((s-1)_i\big) \\
+ \tau\sum_{s'=0}^{s} \mathrm{H}_{ij}(s,s')u_{;ij}(s') \\
+ \sum_{i=1}^{n}\sum_{j=1}^{i}\tau\sum_{s'=0}^{s-1}\mathrm{A}'_{ij}(s,s')\tilde{v}_{;\bar{i}\bar{j}}\left(s'_j\right) = f(s)
\end{aligned}
\tag{10.8}
$$

$$
\begin{aligned}
\frac{v(s_l)-v(s_{l-1})}{\tau} - \sum_{k=1}^{l} P'_{lk}(s)v_{;\bar{l}k}(s_k) + B_{0l}(s)u_{;t\bar{l}}(s) + \sum_{k=1}^{l} B'_{lk}(s)u_{;\bar{l}k}(s) \\
+ \tau\sum_{s'=0}^{s}\sum_{k=1}^{l}\left(\mathrm{P}'_{lk}(s,s')\tilde{v}_{;\bar{l}k}(s'_k) + \mathrm{B}'_{lk}(s,s')u_{;\bar{l}k}(s')\right) = \frac{1}{n}\varphi(s)
\end{aligned}
\tag{10.9}
$$

where $s = 0, \ldots, N$; $l = 1, \ldots, n$.

There is no summation over l in the last equation; dependence on $x \in \Omega_h$ was not written in the above equations for all of the functions included there.

$$
\mathrm{A}'_{ij}(s,s') \equiv (2-\delta_{ij})\mathrm{A}_{ij}(s,s').
$$

δ_{ij} is the Kronecker delta; summation over repeated indices is not meant in this formula. The same sense has the hatch for other coefficients of the grid system.

Let us consider the following boundary conditions for system (10.8) and (10.9):

$$
u(s\tau,x) = 0,\ v(s\tau + l\delta, x) = 0, \tag{10.10}
$$

where $x \in \widetilde{\Omega}_h \backslash \Omega_h$, $s = 0, \ldots, N$ and $l = 1, \ldots, n$.

The initial conditions shall be determined in the form:

$$
u(s\tau,x) = u_0(x),\ \ u_{;t}(0,x) = w_0(x),\ \ v(s\tau + l\delta, x) = v_0(x) \tag{10.11}
$$

where $x \in \Omega_h$, $s \le 0$, and $l = 1, \ldots, n$ for $s < 0$, whereas $l = 0$ for $s = 0$.

We assume that the coefficients under the signs of the sums $\Sigma_{s'=0}^{s}$ in (10.8) and (10.9) are determined for $0 \le s' \le s \le N, x \in \widetilde{\Omega}_h$, and coefficients outside these sums are determined for $0 \le s \le N$ and $x \in \widetilde{\Omega}_h$.

We suppose that the matrix norms of all coefficients in (10.8) and (10.9) outside the sums $\Sigma_{s'=0}^{s}$ are limited along with the matrix norms of their first-order spatial differences (save for A_i) and $H_{ij;t}$, for all nodes of the domain of their definition, by some number M. The coefficients under the sign of a sum are supposed to be limited for $x \in \widetilde{\Omega}_h$ and $y \in \Omega_h$ in the following sense:

$$\tau\sum_{s'=0}^{p}\left(\left|\mathrm{H}_{ij}(s\tau,s'\tau,x)\right|^2+\left|\mathrm{H}_{ij;k}(s\tau,s'\tau,y)\right|^2\right)\leq M$$
$$\tau\sum_{s'=0}^{p}\left(\left|\mathrm{A}_{ij}(s\tau,s'\tau,x)\right|^2+\left|\mathrm{A}_{ij;k}(s\tau,s'\tau,y)\right|^2\right)\leq M$$
$$\tau\sum_{s'=0}^{p}\left(\left|\mathrm{P}_{ij}(s\tau,s'\tau,x)\right|^2+\left|\mathrm{P}_{ij;k}(s\tau,s'\tau,y)\right|^2\right)\leq M \tag{10.12}$$
$$\tau\sum_{s'=0}^{p}\left(\left|\mathrm{B}_{ij}(s\tau,s'\tau,x)\right|^2+\left|\mathrm{B}_{ij;k}(s\tau,s'\tau,y)\right|^2\right)\leq M.$$

Moreover, we suppose that for $i, j = 1, \ldots, n$ and $x \in \Omega_h$

$$\tau\sum_{s'=0}^{p}\left|\mathrm{A}_{ij;t}(p\tau,s'\tau,x)\right|^2\leq M,\quad p=0,\cdots,N-1 \tag{10.13}$$

$$\mid \mathrm{A}_{ij}(s\tau,s\tau,x)\mid + \mid \mathrm{H}_{ij}(s\tau,s\tau,x)\mid\leq M, 0\leq s\leq N. \tag{10.14}$$

Here $\mathrm{A}_{ij;t}$ and $\mathrm{H}_{ij;t}$ are the difference ratios in regard to the first temporal argument:

$$A_{ij;t}(p\tau,s'\tau,x)=\frac{A_{ij}(p\tau+\tau,s'\tau,x)-A_{ij}(p\tau,s'\tau,x)}{\tau}.$$

We suppose the coefficients H_{ij} and P_{ij} are symmetric matrices and their quadratic forms are positive definite, namely for $x\in\widetilde{\Omega}_h$ and $0\leq s\leq N$:

$$\begin{gathered}H_{ij}(s\tau,x)=H_{ij}^{*}(s\tau,x)=H_{ji}(s\tau,x),\\ P_{ij}(s\tau,x)=P_{ji}(s\tau,x)\\ H_{ij}(s\tau,x)w_iw_j\geq\mu w_iw_j\\ P_{ij}(s\tau,x)z_iz_j\geq\mu z_iz_j\end{gathered} \tag{10.15}$$

by some $\mu > 0$. Here $i, j = 1, \ldots, n$; w_i and z_i are arbitrary vectors from R^m и $R^{m'}$, respectively.

These matrices are supposed to be smooth in regard to the temporal and spatial arguments.

The scalar product of vectors with a finite number of components is meant here:

$$zw=\sum_{\alpha=1}^{k}z^{(\alpha)}w^{(\alpha)},\quad z\in R^k,\quad w\in R^k.$$

We denote the grid norms as follows:

$$\sum_x \cdots \equiv \sum_{x\in R_n^h} \cdots$$
$$S_p^q(w) \equiv h^n \sum_{s=p}^{q} \tau \sum_x w(s\tau, x)$$
$$\|w(k)\|_h^2 \equiv h^n \sum_x |w(k\tau, x)|^2 \tag{10.16}$$
$$\|w\|_{\Delta,r}^2 \equiv S_0^r\left(|w|^2\right)$$
$$\|v_{;x}\|_{\Delta,r}^2 \equiv S_0^r(\tilde{v}_{;j}(s_j)\tilde{v}_{;j}(s_j)).$$

Remarks Let us formulate some remarks on the grid operators (finite difference ratio, shift, and summation) that will be used below for investigation of stability of the grid approximation of the initial boundary value problem of the SDSE under consideration.

The finite difference analog of the derivative of the product of two arbitrary grid functions w and z:

$$[wz]_{;i} \equiv \left[w^{+i}z^{+i} - wz\right]/h = \left[w^{+i}z^{+i} - w^{+i}z + w^{+i}z - wz\right]/h = w^{+i}\, z_{;i} + w_{;i}\, z$$
$$= w_{;i}\, z + w^{+i}\, z_{;i}.$$

Then

$$[wz]_{;i} = w_{;i}\, z + w^{+i}\, z_{;i}.$$

This is nearly the formula of Newton–Leibniz, save for the operator of shift w^{+i}. This is the main reason why the analysis of grid functions is so unwieldy as compared with functions of a continuous argument, but the infinitesimal values are absent in the grid analysis.

Using the shift under the sign of $[\ ]_{;i}$ it is possible to avoid the shift above w outside $[\ ]_{;i}$:

$$\left[w^{-i}z\right]_{;i} = w_{;\bar{i}}z + w\, z_{;i}$$

because $[w^{-\,i}];_{\,i} = w_{;\bar{i}}$.

The Newton–Leibniz formula for the derivative of the product of two functions will be used in the form

$$w_{;i}\, z = [wz]_{;i} - w^{+i}\, z_{;i}$$

$$w_{;\bar{i}}z = \left[w^{-i}z\right]_{;i} - w\, z_{;i}.$$

The finite difference "derivative" (i.e., the finite difference ratio) has been changed from w to z, but additional terms $[w\, z]_{;i}$ and $[w^{-\,i}z]_{;i}$ must be added.

Below, similar replacements will be applied along with the summation over R_h^n for the case when the *finite* functions are considered, i.e., the functions that are zero outside some definite domain. Then

$$\sum_x [wz]_{;i} = 0, \quad \sum_x \left[w^{-i}z\right]_{;i} = 0$$
$$\sum_x w_{;i}\, z = \; -\sum_x w^{+i} z_{;i}$$
$$\sum_x w_{;\bar{i}}z = \; -\sum_x wz_{;i}.$$

In this case the *replacement of the finite difference derivative from one function to another* is not connected with arising of additional terms. Surely, it is the known grid formula of summation by parts (it is similar to integration by parts for functions of a continuous argument).

In the case of a finite function, the result of summation over the whole grid does not depend on the spatial shift of its argument.

10.2.3 Stability of the Grid Approximation

Theorem Let conditions (10.12)–(10.15) be satisfied. Then there are the numbers τ_0, k_0, C_0 (they are determined by the above data M, n, μ) such that for $0 < \tau \le \tau_0$ and $0 < \tau/h \le k_0$ the solution of the problem (10.8)–(10.11) exists and is unique, and satisfies the inequality

$$\begin{aligned}&\left\|u_{;t}(r)\right\|_h^2 + \left\|u_{;x}(r)\right\|_h^2 + \|u(r)\|_h^2 + \|v(r)\|_h^2 + \left\|v_{;x}\right\|_{\Delta,r}^2 \\ &\quad \le C_0\left(\|u_0\|_h^2 + \left\|u_{0;x}\right\|_h^2 + \|w_0\|_h^2 + \|v_0\|_h^2 + \|f\|_{\Delta,r}^2 + \|\varphi\|_{\Delta,r}^2\right), \quad 1 \le r \le N.\end{aligned} \tag{10.17}$$

Proof Let us suppose that a solution $u(s)$, $v(s_r)$ of the problem (10.8)–(10.11) exists. Let us continue the coefficients and the solution by zero for $x \in R_h^n/\widetilde{\Omega}_h$, $0 \le s' \le s \le N$. Then we consider Eq. (10.8) written for the node ($s\tau$, x) and multiply both parts of this equation by the function

$\bar{u}_{;t}$, where $\bar{u} \equiv u + u^-$.

The scalar multiplication in R^m is meant here. Because of continuation of the grid functions, the equality obtained is true for all nodes ($s\tau$, x) by $1 \le s \le N$ and $x \in R_h^n$.

Let us estimate the terms included in it. Because of (10.11), we have for $y_0(s) \equiv u_{;t\bar{t}}(s)\bar{u}_{;t}(s)$

$$S_1^r(y_0) = \left\|u_{;t}(r)\right\|_h^2 - \|w_0\|_h^2.$$

With the help of formula $w^+ = w + \tau w_{;t}$ by $w = u_{;\bar{i}}$, we shall write the expression $H_{ij}u_{;\bar{j}}u_{;t\bar{i}}$ in the form of the sum shown in the curly brackets in (10.18). Now we get

$$\begin{aligned} y_1(s) &\equiv -H_{ij}u_{;i\bar{j}}u_{;t} \\ &= -\left(H_{ij}u_{;\bar{j}}u_{;t}^{-i}\right)_{;i} + \frac{1}{2}H_{ij}u_{;\bar{j}}u_{;t\bar{i}} + H_{ij;i}u_{;\bar{j}}^{+i}u_{;t} \\ &+\frac{1}{2}\Big\{\left(H_{ij}u_{;\bar{j}}u_{;\bar{i}}\right)_{;t} - H_{ij}u_{;t\bar{j}}u_{;\bar{i}} - \tau H_{ij}u_{;t\bar{j}}u_{;t\bar{i}} \\ &-H_{ij;t}\left(u_{;\bar{j}} + \frac{\tau}{h}\left(u_{;t} - u_{;t}^{-j}\right)\right)\left(u_{;\bar{i}} + \frac{\tau}{h}\left(u_{;t} - u_{;t}^{-i}\right)\right)\Big\}. \end{aligned} \tag{10.18}$$

Because of the symmetry (10.15) of matrices H_{ij}, the sum of the second term of the right part and the second term in curly brackets gives zero.

Expression $y_2(s) \equiv -H_{ij}u_{;i\bar{j}}u_{;t}^{-}$ may be written in the analogical form with replacement of $u_{;t}$ by $u_{;\bar{t}}$ and $u_{;\bar{j}}^{+}$ by $u_{;\bar{j}}^{-}$. Now the sign "+" will be before the third term in curly brackets in (10.18):

$$\tau H_{ij}u_{;\bar{t}\bar{j}}u_{;\bar{t}\bar{i}} = \tau H_{ij}^{-}u_{;\bar{t}\bar{i}}u_{\bar{t}\bar{j}} + \frac{\tau^2}{h^2}H_{ij;\bar{t}}\left(u_{;\bar{t}} - u_{;\bar{t}}^{-j}\right)\left(u_{;\bar{t}} - u_{;\bar{t}}^{-i}\right)$$

because the formula $w_{;t}^{-} = w - \tau w_{;\bar{t}}$ with $w = u_{;\bar{i}}$ is used in this case. Therefore

$$S_1^r\left(-\frac{\tau}{2}H_{ij}u_{;t\bar{j}}u_{;t\bar{i}} + \frac{\tau}{2}H_{ij}u_{;\bar{t}\bar{j}}u_{;\bar{t}\bar{i}}\right) \geq -\frac{\tau^2}{2}\sum_x H_{ij}(r)u_{;t\bar{j}}(r)u_{;t\bar{i}}(r) - CU_r$$

where

$$\begin{aligned} C &\equiv C(M, n, \mu, T, \tau/h) = \text{const} > 0 \\ U_r &\equiv \left\|u_{;x}\right\|_{\Delta,r}^2 + \left\|u_{;t}\right\|_{\Delta,r}^2 + \|u\|_{\Delta,r}^2 + \|u_{0x}\|_h^2 + \|w_0\|_h^2 + \|u_0\|_h^2. \end{aligned}$$

The sum $\Sigma_x \cdots$ does not depend on the shift of the spatial argument of the grid function under the sign of this sum, in particular,

$$\Sigma_x(\cdots)_{;x} = 0.$$

Here and below, $(\cdots)_{;x}$ denotes a sum of difference spatial ratios of finite in x grid functions (a finite function becomes zero outside some finite domain).

Besides, we take into account everywhere that the estimation

$$\left|H_{ij;i}u^{+i}_{;j}u_{;t}\right| \le C\left(\left|u^{+i}_{;j}\right|^2 + \left|u_{;t}\right|^2\right), \quad 1 \le s \le N$$

is true not only for $x \in \Omega_h$, where $|H_{ij;t}| \le M$ by the hypothesis of the theorem, but for every $x \in R^n_h$ due to (10.10) and continuation of the grid functions described above. Because the quadratic form $H_{ij}w_iw_j$ is positive definite (see (10.15)) we obtain from (10.18) by $1 \le r \le N$

$$S^r_1(y_1+y_2) \ge \frac{\mu}{2}\left(\left\|u_{;x}(r)\right\|^2_h + \left\|u_{;x}(r+1)\right\|^2_h\right) - \frac{2\tau^2Mn^2}{h^2}\left\|u_{;t}(r)\right\|^2_h - C(U_r+F_r)$$

$$F_r \equiv \|f\|^2_{\Delta,r} + \|\varphi\|^2_{\Delta,r} + \left\|u_{0;x}\right\|^2_h + \|w_0\|^2_h + \|v_0\|^2_h.$$

Let us denote by $y_3(s)$ the scalar product of the last term of the left part of (10.8) and vector $\bar{u}_{;t}$.

The sum $\tau\sum_{s'=0}^{s-1}\cdots$, included in this expression, will be written by us as (the summation over repeated indices i and j is not meant there)

$$\begin{aligned}
&\tau\sum_{s'=0}^{s-1}\left(A'_{ij}(s,s')\tilde{v}^{-i}_{;j}\left(s'_j\right)\bar{u}_{;t}(s)\right)_{;i} - \tau\sum_{s'=0}^{s-1}A'_{ij;i}(s,s')\tilde{v}_{;j}\left(s'_j\right)\bar{u}_{;t}(s)\\
&\quad-\left(\tau\sum_{s'=0}^{s-1}A'^{+i}_{ij}(s-1,s')\tilde{v}_{;j}\left(s'_j\right)\bar{u}_{;i}(s)\right)_{;t} + A'^{+i}_{ij}(s,s)\tilde{v}_{;j}\left(s'_j\right)\bar{u}_{;i}(s+1)\\
&\quad+\tau\sum_{s'=0}^{s-1}A'^{+i}_{ij;t}(s-1,s')\tilde{v}_{;j}\left(s'_j\right)\bar{u}_{;i}(s+1).
\end{aligned}\tag{10.19}$$

Hence, using the inequality

$$ab \le \varepsilon|a|^2 + \frac{1}{4\varepsilon}|b|^2, \qquad a,b \in R^k, \qquad k \ge 1, \qquad \varepsilon \ge 0 \tag{10.20}$$

by $\varepsilon = 1$ and the inequality of Cauchy for sums over s', we receive

$$\begin{aligned}
y_3(s) \ge{}& (\cdots)_{;x} - \sum_{i=1}^{n}\sum_{j=1}^{i}(\cdots)_{;t}\\
&-\sum_{i,j=1}^{n}\left\{\left(\tau\sum_{s'=0}^{s-1}\left|A_{ij;i}(s,s')\right|^2 + \tau\sum_{s'=0}^{s-1}\left|A_{ij;t}(s-1,s')\right|^2 + \left|A'^{+i}_{ij}(s,s)\right|\right)\right.\\
&\left.\times\left(\tau\sum_{s'=0}^{s-1}\left|\tilde{v}_{;j}\left(s'_j\right)\right|^2 + \left|\bar{u}_{;t}(s)\right|^2 + \left|\bar{u}_{;i}(s+1)\right|^2 + \left|\tilde{v}_{;j}(s_j)\right|^2\right)\right\}.
\end{aligned}\tag{10.21}$$

Here the third term in (10.19) is denoted by $(\cdots)_{;t}$. Considering the initial conditions and formula,

$$u_{;i}(s)\big|_{s=1} = u_{0;i} + \frac{\tau}{h}\left(w_0^{+i} - w_0\right),$$

we write the sum S_1^r for the appropriate term in (10.21) in the form

$$\sum_{i=1}^{n}\sum_{j=1}^{i}\left\{h^n\tau\sum_x\left(\sum_{s'=0}^{r}\mathrm{A}'^{+i}_{ij}(r,s')\tilde{v}_{;j}\left(s'_j\right)\right)(u_{;i}(r)+u_{;i}(r+1))\right.$$
$$\left.-h^n\frac{\tau}{h}\sum_x\mathrm{A}'^{+i}_{ij}(0,0)\left(\tilde{v}\left(0+j\delta,x^{+j}\right)-\tilde{v}(0+j\delta,x)\right)\left(2u_{0;i}+\frac{\tau}{h}\left(w_0^{+i}-w_0\right)\right)\right\}.$$

The module of this expression will be estimated from the above with the help of inequality (10.20) with an arbitrary $\varepsilon \geq 0$ before $(u_{;i}(r) + u_{;i}(r + 1))$; the value of ε will be chosen later. Then we receive from (10.21) and conditions (10.13) and (10.14)

$$S_1^r(y_3) \geq -C(1+\varepsilon^{-1})\left(\|\tilde{v}_{;x}\|^2_{\Delta,r} + U_r + F_r\right) - \varepsilon\left(\|u_{;x}(r+1)\|_h^2 + \|u_{;x}(r)\|_h^2\right) - \varepsilon V$$
$$V = \sum_{j=1}^{n}h^n\sum_x|v(0+j\delta,x)|^2.$$

The sums S_1^r of scalar products with $\bar{u}_{;t}$ of other terms of the left part of (10.8) may be estimated from below, similarly to the considered sums, through the terms of the left part of the last inequality but without εV.

So, we get the following inequality from (10.8) and the above estimations of S_1^r (y_k):

$$\left(1-2n^2M\frac{\tau^2}{h^2}\right)\|u_{;t}(r)\|_h^2 + \left(\frac{\mu}{2}-\varepsilon\right)\left(\|u_{;x}(r+1)\|_h^2 + \|u_{;x}(r)\|_h^2\right) + \|u(r)\|_h^2$$
$$\leq C(1+\varepsilon^{-1})\left(\|\tilde{v}_{;x}\|^2_{\Delta,r} + U_r + F_r\right) + \varepsilon V \tag{10.22}$$

by $1 \leq r \leq N$ and $\varepsilon \geq 0$.

Now we go over to Eq. (10.9) and multiply both parts of it by $\tilde{v}(s_l)$; the scalar multiplication in $R^{m'}$ is applied. Let us "replace" (see our remarks at the end of the previous section) the finite difference ratio in regard to x_l on $\tilde{v}(s_l)$. Then we shall write every term as a sum of expressions of the form $(\cdots)_{;x}$ and products, including a spatial difference ratio of not higher than the first order of functions u, v and coefficients. We choose the order of application of the operator of difference ratio and shifts of arguments under the sign $(\cdots)_{;x}$ so that the function $\tilde{v}(s_l)$ will be included in the mentioned products without a shift of the spatial argument x.

Using inequality (10.20) with ε before $\tilde{v}(s_l)$ and taking account that coefficients are limited along with their finite difference ratios (see conditions (10.13) and (10.14)) we receive the following estimation by $\varepsilon \geq 0$ and $l = 1, \ldots, n$:

$$\begin{aligned}
&\frac{v^2(s_l) - v^2(s_{l-1})}{\tau} + \sum_{k=1}^{l} P'_{lk}(s)\tilde{v}_{;k}(s_k)\tilde{v}_{;l}(s_l) \\
&\leq (\cdots)_{;x} + C\Bigg\{ \varepsilon\tilde{v}_{;p}(s_p)\tilde{v}_{;p}(s_p) + \left(1 + \frac{1}{\varepsilon}\right)\Bigg[v(s_p)v(s_p) + v^2(s) + u^2_{;t}(s) \\
&+u_{;p}(s)u_{;p}(s) + \tau\sum_{s'=0}^{s}\left(\tilde{v}_{;p}\left(s'_p\right)\tilde{v}_{;p}\left(s'_p\right) + u_{;p}(s')u_{;p}(s')\right)\Bigg] + \varphi^2(s)\Bigg\}.
\end{aligned} \tag{10.23}$$

Here l is fixed and summation over repeated index p from 1 to n is meant. Let us note that the inequality

$$|\tilde{v}(s_l)|^2 \leq C\left(v(s_p)v(s_p) + |v(s)|^2\right)$$

was used in the right part. Besides, the shifts of the spatial arguments (regarding the arbitrary fixed $x \in \Omega_h$) of functions $\tilde{v}^2_{;p}(s_p)$, $u^2_{;t}(s)$, and $u^2_{;p}(s)$ were not written out because the result of summation of these functions over $x \in R^n_h$ does not depend on shifts (see our remarks at the end of the previous section). Therefore, shifts may be taken into account by including the additional term in C depending on n only.

Let us sum both parts of (10.23) over l from 1 to n, and then over $x \in R^n_h$, and over s from 0 to r, $1 \leq r \leq n$. Now, using condition (10.15), we get

$$\begin{aligned}
&\|v(r+1)\|^2_h + (\mu - \varepsilon C)\left\|\tilde{v}_{;x}\right\|^2_{\Delta,r} \\
&\leq C\left(1 + \frac{1}{\varepsilon}\right)\left[S^r_0(v(s_p)v(s_p)) + \|v\|^2_{\Delta,r} + \tau\sum_{s=0}^{r}\left\|v_{;x}\right\|^2_{\Delta,s} + U_r + F_r\right].
\end{aligned} \tag{10.24}$$

To exclude V from (10.22) and S^r_0 from (10.24), we shall choose $\varepsilon = 1$ in (10.23) and sum both parts over l from 1 to p, where $1 \leq p \leq n$, and then sum over p from 1 to n. Because function v has no shift of the spatial argument in the resulting inequality, we have

$$\begin{aligned}
&(1 - C\tau)v(s_p)v(s_p) \leq (\Lambda)_{;x} + C(1+\tau)v^2(s) \\
&+\tau C\Bigg\{ \tilde{v}_{;p}(s_p)\tilde{v}_{;p}(s_p) + u_{;p}(s)u_{;p}(s) + u^2_{,t}(s) \\
&+\tau\sum_{s'=0}^{s}\left(\tilde{v}_{;p}\left(s'_p\right)\tilde{v}_{;p}\left(s'_p\right) + u_{;p}(s')u_{;p}(s')\right) + \varphi^2(s)\Bigg\}.
\end{aligned} \tag{10.25}$$

By μ, n, M, T fixed by the conditions of the theorem we shall choose $k_0 \geq 0$ and $\tau_1 > 0$ small enough and the inequalities will be satisfied:

$$1 - 2n^2 M k_0^2 \geq \frac{1}{2}, \qquad \tau_1 C(M, n, \mu, T, k_0) \leq \frac{1}{2}. \tag{10.26}$$

From (10.25) by $0 < \tau < \tau_1$ and $0 < \tau/h < k_0$ we find

$$S_0^r\big(v(s_p)v(s_p)\big) \leq \tau_1 C_1 \left\|\tilde{v}_{;x}\right\|^2_{\Delta,r} + C_1(1+\tau_1)\Big(U_r + F_r + \|v\|^2_{\Delta,r}\Big)$$

where $C \equiv C(M, n, \mu, T, k_0)$. Summing (10.25) over $x \in R_h^n$ by $s = 0$ gives

$$V \leq (1+\tau_1)C_1\Big(\left\|\tilde{v}_{;x}\right\|^2_{\Delta,0} + F_0\Big).$$

With the help of the last estimations S_0^r and V we exclude the values of the function v with the fractional (regarding τ) argument from (10.22) and (10.24). Then we multiply (10.24) by $\gamma > 0$ (it will be chosen later) and add inequality (10.22). Because of (10.26), we find

$$\begin{aligned} &\|u_{,t}(r)\|_h^2 + \left(\frac{\mu}{2} - \varepsilon\right)\left(\left\|u_{;x}(r+1)\right\|_h^2 + \left\|u_{;x}(r)\right\|_h^2\right) + \|u(r)\|_h^2 + \gamma\|v(r+1)\|_h^2 \\ &\quad + \left\{\gamma(\mu - \varepsilon C_1 - (1+\varepsilon^{-1})C_1\tau_1) - C_1(\varepsilon + \varepsilon\tau_1 + \varepsilon^{-1})\right\} \left\|\tilde{v}_{;x}\right\|^2_{\Delta,r} \\ &\quad \leq (1+\gamma)(1+\varepsilon^{-1})C_1(1+\tau_1)\left(\|v\|^2_{\Delta,r} + \tau\sum_{s=0}^{r}\left\|\tilde{v}_{;x}\right\|^2_{\Delta,s} + U_r + F_r\right) \end{aligned} \tag{10.27}$$

for $0 < \tau < \tau_1$ and $0 < \tau/h < k_0$.

Let us choose $\varepsilon = \varepsilon_0$ small enough to get inequalities

$$\frac{\mu}{4} - \varepsilon_0 C(M, n, \mu, T, k_0) \geq 0, \qquad \frac{\mu}{4} - 2\varepsilon_0 \geq 0. \tag{10.28}$$

Now we choose $\tau_0 \leq \tau_1$ small enough, and then the following inequality will be true, along with (10.27):

$$¼\,\mu - \tau_0 C(M, n, \mu, T, k_0)(1 + 1/\varepsilon_0) \geq 0. \tag{10.29}$$

Finally, we shall choose $\gamma_0 \geq 0$ large enough to get the inequality

$$¼\gamma_0 - C(M, n, \mu, T, k_0)(\varepsilon_0 + 1/\varepsilon_0 + \varepsilon_0\tau_0) \geq 0. \tag{10.30}$$

After choosing the numbers ε_0, v_0, and k_0 in accordance with (10.26) and (10.28)–(10.30), we exclude $\|v\|_{\Delta,r} U_r$ and $\tau\sum_{s=0}^{r}\left\|\tilde{v}_{;x}\right\|^2_{\Delta,s}$ from (10.27) with the help of the discrete analog of the lemma of Gronwall (Yang 1988). Thus, we receive (10.17). The single-valued solvability of the problem (10.8)–(10.11) follows from this inequality. *End of the proof.*

10.3 Galerkin Method for the 3D Nonlinear System of PDEs of Magneto-thermo-elasticity

10.3.1 Initial Boundary Value Problem

We consider the nonlinear system of PDEs of the dynamical coupled theory of magneto-thermo-elasticity (Knopoff 1955; Maugin 2013):

$$\rho\ddot{u} = \mathrm{Div}\ \hat{\tau} - \mu_e[h \times \mathrm{rot}\ h] + F \tag{10.31}$$

$$\sigma\mu_e\dot{h} = \Delta h + \sigma\mu_e \mathrm{rot}\left[\dot{u} \times h\right] \tag{10.32}$$

$$\rho c_V\dot{\vartheta} = k\Delta\vartheta - \vartheta_0\gamma_0\,\mathrm{div}\,\dot{u} + Q. \tag{10.33}$$

The stress tensor is determined by the law of Hook–Duhamel–Neumann:

$$\tau_{\alpha\beta} = \mu\cdot\left(u_{\alpha,\beta} + u_{\beta,\alpha}\right) + \delta_{\alpha\beta}\cdot(\lambda\ \mathrm{div}u - \gamma_0\vartheta). \tag{10.34}$$

Here $x = (x_1, x_2, x_3)$ are the spatial coordinates, $u = (u_1, u_2, u_3)$ is the vector of elastic displacements, $u_{\alpha,\beta} \equiv \partial u_\alpha/\partial x_\beta$, $h = (h_1, h_2, h_3)$ is the vector of magnetic intensity, ϑ is disturbance of the temperature regarding the reference temperature ϑ_0, and F and Q are the densities of the mass forces and the heat sources; other values (i.e., μ, λ, and so on) are the given characteristics of the medium; the vector product of the three component vectors is denoted with [].

The system (10.31)–(10.34)—based on the conservation laws (theory of elasticity), the electrodynamics of slowly moving media, and thermodynamics—describes the coupled effects in deformable bodies with electrical and heat conductivity for the case of negligible displacement currents (as compared with conductive currents), the negligible volume charge, and joule heat production. The system (10.31)–(10.34) belongs to the SDSE type (see Chap. 9) and may be applied for investigations of disturbances of the EM field during seismic activations.

The energy estimations (Ladyjenskaya et al. 1967) of the solutions of the initial boundary value problems of the SDSEs of the thermo-mechanical coupling, the correctness of these problems in the sense of Hadamard, and the stability and convergence of their finite difference approximations were considered by Novik (1969) and by Mikhaylovskaya and Novik (1979). The SDSE of the nonlinear dynamic theory of magneto-thermo-elasticity for the two-dimensional case was considered by Novik (1993).

Below, with the help of the energy estimations, we have proved, in the three-dimensional case, that for the Galerkin approximation of the initial boundary value problem (with the Dirichlet boundary conditions) of the nonlinear magneto-thermo-elasticity system (10.31)–(10.34), the solution exists in an arbitrary finite time interval [0, *T*] (Novik 1995). In addition, it is shown in this work that the functional energy norm of the Galerkin approximate solution—by every $0 \leq t \leq T$ and arbitrary number *N* of the functions of the Galerkin basis—is bounded by the value determined by the physical characteristics of the material, the Sobolev norms of the considered fields by $t = 0$, and the intensity of the sources of the fields. Let us note that the squared functional energy norm majorizes the summary energy of the elastic, magnetic, and thermal fields.

We shall consider the following initial and boundary conditions for the system (10.31)–(10.34):

$$\begin{aligned} u(t,x)|_{t=0} &= u_0(x) \\ \dot{u}\,(t,x)|_{t=0} &= u_1(x) \\ h(t,x)|_{t=0} &= h_0(x) \\ \vartheta(t,x)|_{t=0} &= \vartheta^{(0)}(x) \\ u(t,x)|_{\partial\Omega} &= 0 \\ h(t,x)|_{\partial\Omega} &= 0 \\ \vartheta(t,x)|_{\partial\Omega} &= 0. \end{aligned} \tag{10.35}$$

Here Ω is a finite domain of the three-dimensional space R^3; $x \in \Omega$, $t \in [0, T]$. We will apply the following notations for the scalar product and norms:

$$\begin{aligned} (w,z) = (w(t), z(t)) &\equiv \textstyle\int_\Omega w_\alpha(t,x) z_\alpha(t,x)dx \\ \|w(t)\| &= (w(t), w(t))^{1/2} \\ \|w_x(t)\| &= \left(w_{\alpha,\beta}(t),\ w_{\alpha,\beta}(t)\right)^{1/2} \\ \|w(t)\|_1 &= \|w(t)\| + \|w_x(t)\|. \end{aligned}$$

10.3.2 Theorem of Existence of the Galerkin Approximation

To construct the Galerkin approximation to the solution of the problem (10.31)–(10.35) we apply the system of functions $g_p(x) \in H_0^1(\Omega)$: $(g_k, g_l) = \delta_{kl}$, $\delta_{kl} = 0$ for $k \neq l$, $\delta_{kk} = 1$, summation on *k* is absent here, the linear span of the functions $g_p(x)$, $p = 1, 2, 3, \ldots$ is supposed to be dense in $H_0^1(\Omega)$; the last designates the Sobolev space of functions $f(x)$: $f(x)$, $\nabla f(x) \in L_2(\Omega)$ (every component is meant), $f(x)|_{\partial\Omega} = 0$; the derivatives of the functions from $L_2(\Omega)$ and boundary values are understood in the sense of Sobolev (Sobolev and Vaskevich 1997; Adams and Fournier 2003; Leony 2009).

The magneto-thermo-elasticity system (10.31)–(10.33) is written using the Galerkin approximation in the following form:

$$\rho\left(\ddot{u}_\alpha, g_l^N\right) = -\left(\tau_{\alpha\beta}^N,\ g_{l,\beta}\right) - \mu_e \cdot \left(\left[h^N \times \mathrm{rot} h^N\right]_\alpha,\ g_l\right) + \left(F_\alpha, g_l\right) \tag{10.36}$$

$$\sigma\mu_e \cdot \left(\dot{h}_\alpha^N, g_l\right) = -\left(h_{\alpha,\beta}^N,\ g_{l,\beta}\right) + \sigma\mu_e\left(\mathrm{rot}\left[\dot{u}^N \times h^N\right]_\alpha,\ g_l\right) \tag{10.37}$$

$$\rho c_V\left(\dot{\vartheta}^N, g_l\right) = -k \cdot \left(\vartheta_{,\beta}^N,\ g_{l,\beta}\right) + \vartheta_0\gamma_0 \cdot \left(\dot{u}_\beta^N,\ g_{l,\beta}\right) + \left(Q, g_l\right). \tag{10.38}$$

Here α =1, 2, 3; $l = 1, \ldots, N$; and the Nth Galerkin approximation of the solution of the problem under consideration is denoted by

$$\begin{aligned} u_\alpha^N &= a_{al}^N(t)g_l(x) \\ h_\alpha^N &= b_{al}^N(t)g_l(x) \\ \vartheta^N &= c_l^N(t)g_l(x) \\ u^N &= \left(u_1^N, u_2^N, u_3^N\right) \\ h^N &= \left(h_1^N, h_2^N, h_3^N\right). \end{aligned}$$

The component with the number α of the vector w is denoted by w_α or $[w]_\alpha$, if w is a vector valued expression.

The repeating Latin index denotes summation from 1 to N. The Nth approximation of the stress and deformation tensors is written using the following formulas:

$$\begin{aligned} \tau_{\alpha\beta}^N &= 2\mu e_{\alpha\beta}^N + \left(\lambda e^N - \gamma_0\vartheta^N\right)\delta_{\alpha\beta}\ , \\ e_{\alpha\beta}^N &= \frac{u_{\alpha,\beta}^N + u_{\beta,\alpha}^N}{2}, \\ e^N &= e_{\alpha,\alpha}^N\ . \end{aligned}$$

The initial conditions for the Galerkin decompositions of the approximate solution are chosen in the form

$$a_{\beta l}^N(t)\Big|_{t=0} = \alpha_{\beta l}^N\ , \qquad \dot{a}_{\beta l}^N(t)\Big|_{t=0} = \left(u_{1\beta}, g_l\right)\ , \tag{10.39}$$

$$b_{\beta l}(t)\big|_{t=0} = \left(h_{0\beta}, g_l\right), \qquad c_l(t)\big|_{t=0} = \left(\vartheta^{(0)}, g_l\right). \tag{10.40}$$

Here β =1, 2, 3; $l = 1, \ldots, N$; and initial values $\alpha_{\beta l}^N$ of the coefficient $a_{\beta l}^N(t)$ were chosen to provide the convergence

$$u_\beta^N(0,x) = \alpha_{\beta l}^N g_l(x) \to u_{0\beta}(x) \quad \text{in } H^1(\Omega), \tag{10.41}$$

in accordance with the mentioned properties of the system $g_l(x)$; by these conditions for an arbitrary N the inequality is true: $\|u(0)\|_1 \leq C\|u_0\|_1$.

Theorem Let us suppose that $\partial\Omega \in C^{(1)}$, $u_0(x) \in \left(H_0^1\right)^3$, $u_1(x) \in (L_2(\Omega))^3$, $h_0(x) \in (L_2(\Omega))^3$, and $\vartheta^{(0)}(x) \in L_2(\Omega)$. Then the solution $a_{al}^N(t)$, $b_{al}^N(t)$, $c_l^N(t)$ of the Cauchy problem (10.36)–(10.41) for the Galerkin system of equations exists at an arbitrary time interval $t \in [0, T]$ and obeys the inequality

$$\begin{aligned}
&\|\dot{u}^N(t)\| + \|u^N(t)\|_1 + \|h^N(t)\| + \|\vartheta^N(t)\| + \textstyle\int_0^t \left(\|h_x^N(t')\| + \|\nabla\vartheta^N(t')\|\right)dt' \\
&\leq C \times \left(\|u_0\|_1 + \|u_1\| + \|h_0\| + \|\vartheta^{(0)}\| + \sqrt{\textstyle\int_0^t \left(\|F(t')\|^2 + \|Q(t')\|^2\right)dt'}\right)
\end{aligned}$$

uniformly in regard to N; $C = C(\Omega, T, \mu, \lambda, \rho, c_V, \sigma, \mu_e, k, \gamma_0, \vartheta_0)$ (Novik 1994a, 1995).

Proof Let us increase every term of the Galerkin equation of motion (10.36) on $\sigma\dot{a}_{al'}^N(t)$, for the equation of magnetic diffusion (10.36) on $b_{al'}^N(t)$, and for the equation of heat conduction on $c_{l'}^N(t)$. Let us summarize the transformed equations over l from 1 to N and estimate separate terms of the resulting equality. With application of the zero boundary conditions and symmetry of the tensor of deformation, we will consider the Galerkin analog of the power of elastic forces in the body Ω at the moment t:

$$\begin{aligned}
-P^N(t) &\equiv \left(\tau_{\alpha\beta}^N(t),\ \dot{u}_{\alpha,\beta}^N(t)\right) \\
&= 2\mu\left(e_{\alpha\beta}^N,\ \left(\dot{u}_{\alpha,\beta}^N + \dot{u}_{\beta,\alpha}^N\right)/2\right) + \lambda\left(u_{\gamma,\gamma}^N,\ \delta_{\alpha\beta}\dot{u}_{\alpha,\beta}^N\right) + \gamma_0\left(\vartheta_{,\beta}^N,\ \delta_{\alpha\beta}\dot{u}_\alpha^N\right) \\
&= \mu\partial_t\left(e_{\alpha\beta}^N(t), e_{\alpha\beta}^N(t)\right) + \frac{\lambda}{2}\partial_t\|\mathrm{div}u^N(t)\|^2 + \gamma_0\left(\nabla\vartheta^N(t), \dot{u}^N(t)\right).
\end{aligned}$$

Now let us consider power $P_M^N(t)$ of the magnetic forces:

$$\begin{aligned}
\frac{1}{\mu_e}P_M^N(t) &\equiv -\left(h^N(t) \times \mathrm{rot}h^N(t),\ \dot{u}^N(t)\right) \\
&= -\left(\dot{u}^N(t), h^N(t) \times \mathrm{rot}h^N(t)\right) \\
&= -\left(\dot{u}^N(t) \times h^N(t),\ \mathrm{rot}h^N(t)\right).
\end{aligned}$$

Here the formula was used where the dot denotes the scalar product and [] denotes the vector product.

$$A \cdot [B \times C] = [A \times B] \cdot C$$

by $A = \dot{u}^N(t)$, $B = h^N(t)$ и $C = \mathrm{rot}h^N(t)$.

Let us transform the term of the Galerkin equation of magnetic diffusion (10.37) including the velocity of displacements:

$$\frac{1}{\sigma\mu_e} M^N(t) \equiv \left(\mathrm{rot}\left[\dot{u}^N(t) \times h^N(t)\right],\ h^N(t)\right) = \left(\dot{u}^N(t) \times h^N(t),\ \mathrm{rot}h^N(t)\right)$$

with the help of the formula

$$\mathrm{div}[R \times S] = S \cdot \mathrm{rot}R - R \cdot \mathrm{rot}S,$$

which is true for an arbitrary vector field $R, S \in (H^1(\Omega))^3$. Then

$$\int_\Omega \mathrm{div}[R \times S] dx = \int_{\partial\Omega} [R \times S] \cdot n_{\partial\Omega}\ ds = 0$$

if one of the vectors is zero at $\partial\Omega$; $n_{\partial\Omega}$ is the normal to the boundary.

Therefore, for similar vectors

$$(S, \mathrm{rot}R) = (R, \mathrm{rot}S).$$

We choose

$$S = \dot{u}^N(t) \times h^N(t), \qquad R = h^N(t),$$

and obtain the above expression for $M^N(t)$. So,

$$\sigma P_M^N(t) = M^N(t) = 0.$$

Now from equations of motion, magnetic field, and heat conduction we obtain for arbitrary values of $t > 0$ и $\varepsilon > 0$:

$$\begin{aligned}
&\frac{\sigma\rho}{2}\partial_t\left\|\dot{u}^N(t)\right\|^2 + \sigma\mu\partial_t\left(e_{\alpha\beta}^N(t),\ e_{\alpha\beta}^N(t)\right) + \frac{\sigma\lambda}{2}\partial_t\left\|\mathrm{div}u^N(t)\right\|^2 + \frac{\sigma\mu_e}{2}\partial_t\left\|h^N(t)\right\|^2 \\
&\quad + \frac{\rho c_V}{2}\partial_t\left\|\vartheta^N(t)\right\|^2 + \left\|h_x^N(t)\right\|^2 + k\left\|\nabla\vartheta^N(t)\right\|^2 \\
&\le 2\varepsilon\left\|\nabla\vartheta^N(t)\right\|^2 + \frac{(\gamma_0\sigma)^2}{4\varepsilon}\left\|\dot{u}^N(t)\right\|^2 + \frac{(\vartheta_0\gamma_0)^2}{4\varepsilon}\left\|\dot{u}^N(t)\right\|^2 + \frac{1}{2}\|F(t)\|^2 \\
&\quad + \frac{\sigma^2}{2}\left\|\dot{u}^N(t)\right\|^2 + \frac{1}{2}\|Q(t)\|^2 + \frac{1}{2}\left\|\vartheta^N(t)\right\|^2.
\end{aligned}$$

We choose $\varepsilon = k/4$ and integrate this inequality termwise over time from 0 up to some arbitrary moment. Then we obtain that norms under the sign of the derivative on time and from the two last terms of the left-hand part and $\|u^N(t)\|^2$ are bounded by the sum of the initial values of the first five norms of the left-hand part, the expression $\|u^N(0)\|^2$ and integrals over time from all terms of the right-hand part, except for the first term. With the help of the inequality of Gronwall (see the *Encyclopedia of Mathematics* and any textbook on differential equations), the norms of unknown functions may be excluded from the right-hand part of this inequality. Then we apply the inequality of Korn (Wang 2003; Beckenbach and Bellman 1975):

$$\left\|u^N(t)\right\|_1^2 \leq C(\Omega) \quad \times \left\{\left(e^N_{\alpha\beta}(t), e^N_{\alpha\beta}(t)\right) + \left\|u^N(t)\right\|\right\},$$

which is true for any $t > 0$. Taking into account that because of the choice of the initial values for $a^N_{\alpha t}(t)$

$$\left(e^N_{\alpha\beta}(0), e^N_{\alpha\beta}(0)\right) \leq C\left(u^N_{\alpha,\beta}(0), u^N_{\alpha,\beta}(0)\right) \leq C\|u_0\|_1^2,$$

we obtain the theorem.

Remarks This proof of the energy correctness of the Galerkin method for the system of equations of magneto-thermo-elasticity may be used for another type of boundary conditions, nonuniform media, magneto-thermo-elastic interaction, and magneto-thermo-hydrodynamic interaction. On the basis of the theoretical results described here, the seismo-EM phenomena were calculated in the initial part of this book, including the seismic excitation of a low-resistivity block, i.e., a typical element of a seismically active lithosphere zone.

10.4 Conclusion

In this chapter, some results on the solvability of problems for the string–diffusion systems of equations (SDSE; see Chap. 9) have been obtained, e.g., in the case of the net approximation of the integro-differential ("memory" of the medium) system with the hyperbolic and parabolic operators of the second order (for other examples, see References).

It has been proved that for the Galerkin approximation of an initial boundary value problem of a nonlinear magneto-thermo-elasticity system, the solution exists on an arbitrary finite time interval [0, *T*]. Moreover, it has been shown that the functional energy norm of the Galerkin approximate solution—by every $0 \leq t \leq T$ and an arbitrary number of the functions of the Galerkin basis—is bounded by the value determined by the physical characteristics of the material, the Sobolev norms of the considered fields by $t = 0$, and the norms of the sources of the fields.

References

Adams, R., & Fournier, J. (2003). *Sobolev spaces* (p. 320). Amsterdam: Elsevier.

Babich, V. M. (2009). *Asymptotic methods in short-wavelength diffraction theory (alpha science series on wave phenomena)*. Oxford: Alpha Science.

Beckenbach, E., & Bellman, R. (1975). *Introduction to inequalities*. New York: Random House.

Egorov, Y. V. (1984). *Linear differential equations of the main type*. Moscow: Nauka (In Russian: Линейные дифференциальные уравнения главного типа. Москва, Наука. 359 с.).

Knopoff, L. (1955). The interaction between elastic wave motion and a magnetic field in electrical conductors. *Journal of Geophysical Research, 60*(4), 441–446.

Ladyjenskaya, O. A., Solonnikov, V. A., & Uraltseva, N. N. (1967). *Linear and quasilinear equations of parabolic type. Translations of mathematical monographs* (Vol. 23). Providence: American Mathematical Society. (Translated from Russian, Moscow: Izdat. Nauka 1964).

Ladyjenskaya, O. A. (1969). *The mathematical theory of viscous incompressible flow*. New York: Gordon and Breach, Science Publications. (Translated from Russian, Moscow: Gos. Izdat. Fiz-Mat Lit. 1961).

Leony, G. (2009). *A first course in Sobolev spaces* (p. 607). Providence: American Mathematical Society.

Maugin, G. (2013). *Continuum mechanics trough the twentieth century*. New York: Springer.

Mikhaylovskaya, I. B., Novik, O. B. (1979). *The Cauchy problem in the class of growing functions for a non-hypoelliptic system of equations which is not hyperbolic*. (In Russian: Михайловская И.Б., Новик О.Б. Задача Коши в классе растущих функций для негипоэллиптической системы уравнений, не являющейся гиперболической //Сибирский математический журнал. Деп. ВИНИТИ. 1979.Ч. 1. № 2104–79. 22 с. Ч. 2. № 2105–79. 22 с.).

Mikhaylovskaya, I. B., & Novik, O. B. (1995). *Net approximation of a conservative–dissipative system of equations*. (In Russian: Михайловская И.Б., Новик О.Б. Сеточное приближение консервативно-диссипативной системы уравнений // «Фундаментальные и прикладные проблемы механики деформируемых сред и конструкций». Программа Госкомвуза РФ по высшему образованию: Науч. труды. Вып. 2. Изд-во Нижегородского гос. ун-та, 1995. Стр. 21–33).

Novik, O. B. (1969). The Cauchy problem for the system of PDEs including the hyperbolic and parabolic operators. *Journal of Computational Mathematics and Mathematical Physics, 1*, 122–136 (In Russian).

Novik, O. B. (1983a). *Interaction of the geothermic field with the field of deformations of a layer*. Geothermic investigations in Middle Asia and Kasahstan. Abstracts of reports of all-union conference, Ashgabat (In Russian).

Novik, O. B. (1983b). *Mathematical description of non-stationary fields of deformations and temperature*. In Proceedings of high educational institutes (Известия вузов) (p. 24). Ser. Geology and Prospecting. Deposited in VINITI. № 6788-83 (In Russian).

Novik, O. B. (1988). Finite difference schemes for the problems of thermo-mechanical effect in the oceanic lithosphere: collection of scientific papers. In *Geothermic investigations on the bottoms of the water areas* (pp. 89–97). Moscow: Nauka (Наука) (In Russian).

Novik, O. B. (1993). *Galerkin method for the 2D system of equations of magneto-thermo-elasticity*. (In Russian: Новик О.Б. Метод Галеркина для двумерной системы уравнений магнито-термо-упругости: Сб. «Фундаментальные и прикладные проблемы механики деформируемых сред и конструкций». Вып. 1. Н. Новгород, 1993. Стр. 117–124).

Novik, O.B. (1994a). *Galerkin method for the 3D non-linear system of equations of magneto-thermo-elasticity*. (In Russian: Новик О.Б. Метод Галеркина для трехмерной нелинейной системы уравнений магнито-термо-упругости. Доклады РАН. 1994. Т. 334. № 2. Стр. 100–102. The Journal is translated and published in English).

Novik, O. B. (1994b). *Seismogenic magneto-thermo-elastic effect*. (In Russian: Новик О.Б. Сейсмогенный магнитотермоупругий эффект. Доклады РАН. 1994б. Т. 338. № 2. Стр. 238–241. The Journal is translated and published in English).

Novik, O. B. (1995). *Interaction of heat-transfer in the active lithosphere zones with other geophysical fields*. (In Russian: Новик О.Б. Взаимодействие теплопереноса в активных зонах литосферы с другими геофизическими полями. Международный симпозиум «Проблемы геотермальной энергии»: Сб. основных докладов. 1995. Стр. 56–73).

Rihtmayer, R. D., & Morton, K. W. (1994). *Difference methods for initial-value problems* (2nd ed.). Malabar: Krieger.

Sobolev, S. L., & Vaskevich, V. (1997). *Theory of cubature formulas*. New York: Springer.

Yang, E. H. (1988). On some new discrete generalizations of Gronwall's inequality. *Journal of Mathematical Analysis and Applications, 129*(2), 505–516.

Wang, L. (2003). On Korn's inequality. *Journal of Computational Mathematics, 21*(3), 321–324.

Chapter 11
A Model Description of Rolling Waves on Water

Abstract In this chapter we investigate, from the mathematical viewpoint, the problem of vertical hydrodynamic wave fronts (shock waves) arising as a result of nonlinear tsunami process development. As the model PDE (partial differential equation) we apply the Burgers equation of turbulence, modified by us with a term without differentiation. For this equation, we consider the vanishing coefficient of viscosity $\varepsilon \to 0$ in the class of self-similar solutions, depending on the linear combination $x = y - ct$ of the spatial (y) and temporal (t) arguments, where $c = \text{const}$ is the coefficient; similar solutions are referred to by us as rolling wave solutions (rw-solutions). These solutions are smooth by $\varepsilon > 0$. The shock rw-solutions of the model PDE with $\varepsilon = 0$ are considered in the same class of the self-similar solutions depending on x but with an analog of the hydrodynamic conditions in the breakpoints, i.e., vertical fronts (known as the conditions on a jump). An arbitrary shock rw-solution $U_0(t, y)$ is approximated with the smooth rw-solutions $U_\varepsilon(t, y)$: $U_\varepsilon(t, y) \to U_0(t, y)$, $\varepsilon \to 0$, according to the rule $\Phi(U_\varepsilon(t, y)) = \Phi\,(U_0(t, y))$, $\varepsilon > 0$, where Φ is the functional determined on $U_\varepsilon(t, y))$, $\varepsilon \geq 0$.

It was shown for the modified Burgers equation that the class of functionals allowing for an arbitrary fixed shock rw-solution to construct the approximating sequence of the smooth rw-solutions obeying this rule is not empty; $U_\varepsilon(t, y)$ is determined uniquely by $\varepsilon > 0$.

Keywords Modified Burgers equation · Vanishing viscosity · Self-similar solutions

11.1 Introduction: The Model Equation with Vanishing Viscosity and Rolling Wave Solutions

In this chapter, we continue the description of the mathematical background of seismo-hydro-electromagnetics. Now we are interested in the nonlinear hydrodynamic component of field interaction. It is rational to consider this component separately because it is the most difficult component from the mathematical point of view. Indeed, there are plenty of unsolved problems with the hydrodynamic

O. Novik et al., *Electromagnetic Geophysical Fields*,
https://doi.org/10.1007/978-3-319-98461-2_11

theory, such as the problem of shock waves, i.e., ones where a sharp change in their characteristics (e.g., jumps in pressure, temperature) in narrow domains of flow are observed. In other words, the smooth (differentiable) dependence of the characteristics of the flow on the coordinates is lost in the course of a nonstationary flow process ("gradient catastrophe"). The question arises: How can we apply the differential equations of hydrodynamics in the absence of differentiability of the characteristics of the flow? Shock waves are absent in the case of a fluid with high viscosity, but these waves are possible if the magnitude of viscosity is small. Usually, the viscosity terms in the hydrodynamic equations include the spatial derivatives of the second order with a coefficient of viscosity before them. Therefore, to describe shock waves, it is necessary to consider hydrodynamic equations with a small coefficient (of viscosity) before the second-order derivatives, e.g., Laplacian ("vanishing viscosity"). Similar terms determine the mathematical type of an equation, and it is possible to discuss the vanishing type. A similar mathematical problem is known in the most complicated part of the hydrodynamic theory, namely in the theory of turbulence. To proceed with the mathematical investigation of hydrodynamics, the model differential equations are applied, which seem to be simpler than the full original equations but, like the full equations, they include the product of the unknown function and its derivative. The vanishing viscosity problem has been investigated in mathematics for some equations by the condition of the fixed initial value of the "viscous solutions" for all values of the viscosity coefficient. (Surely, the solution of the equation with a nonzero coefficient of viscosity is meant here as the viscous solution). But there are cases for which the characteristics of the flow depend on some special combinations of the time (t) and spatial (x) arguments; these flows and corresponding solutions of the equations are referred to as self-similar, e.g., a tsunami wave rolling far from a shore. It will be natural for the self-similar case to consider vanishing viscosity in the class of self-similar solutions. But is it possible? Indeed, in this case, application of the condition of the fixed initial ($t = 0$) value of the "viscous solutions" is impossible just because of the self-similarity of the investigated solutions. Namely, if the initial value of a self-similar solution $u_\varepsilon(x - ct)|_{t=0}$, depending on a linear combination of the time and coordinate, is fixed, then the profiles of all viscous solutions are fixed and approximation of the vertical front of a discontinuous solution (shock wave with $\varepsilon = 0$) by the steep fronts of the viscous solution with the small $\varepsilon > 0$ would be impossible.

Nevertheless, the authors have investigated the vanishing viscosity for the self-similar solutions of the type of a rolling wave (rw-solution; depending on the linear combination t and x; see below) for the model equation of turbulence of Burgers modified by an additional term including the unknown function without derivatives. The way to choose the rw-solution for every positive value of the viscosity coefficient was found and the convergent sequence of rw-solutions was constructed. The rw-solution with the vertical front and zero viscosity coefficient is approximated (by a proper functional norm) by smooth rw-solutions with a decreasing viscosity coefficient and increasing steepness of the fronts.

The equation of Burgers is widely applied for the study of problems of turbulence in different domains of theoretical and mathematical physics, including fluid mechanics, astrophysics, and partial differential equations (PDEs) (Dressler 1949; Landau and Lifshitz 1987; Bec and Khanin 2007).

By some characteristics of an open inclined channel and acting disturbances, which have not been investigated well enough, such waves do arise with distinctive turbulence, periodicity, constant velocity, and considerable concentration of the mass of a fluid within narrow zones near the waves' fronts. Similar waves were referred to as "roll-waves" by Dressler (1949). They generate sharp pulsating stresses in hydro-technic constructions; thus, a decrease in the debit of the flow becomes necessary. Because of the self-similarity considered, we apply the expression of rolling waves.

According to the periodicity and constant velocity of a rolling wave, the periodic self-similar solutions

$$U(t, y) = u(x), x \equiv y - ct$$

of the hydrodynamic equations should be applied for the mathematical description of the rolling wave; here y is the coordinate along the direction of motion of the rolling wave, the velocity of the rolling wave is denoted by c, t is time, and $u(x)$ is the profile of the rolling wave.

Below, only self-similar solutions of this type will be considered (more detailed definitions are formulated in the next section).

This chapter is devoted to a model description of rolling waves with the help of the rw-solutions of the equation

$$\partial U/\partial t + 1/2\partial\left(U^2\right)/\partial y = d(U - c) + \varepsilon\, \partial^2 U/\partial y^2, \ -\infty < d < \infty, \varepsilon \geq 0. \tag{11.1}$$

Formally, Eq. (11.1) differs from the original equation of Burgers by the existence of the first term ("force") in the right part, because $d = 0$ in the original equation of Burgers. Equation (11.1) without the last term, characterizing the influence of viscosity, does not have smooth rw-solutions, but it has piecewise–smooth rw-solutions; see below. The inclusion in Eq. (11.1) of the viscosity term $\varepsilon\partial^2 U/\partial y^2$ results in disappearing of the sharp picks and vertical fronts of the profiles of the rw-solutions. The "smearing" of these fronts is arbitrary small if $\varepsilon > 0$ is small enough (see below).

After description of the problem in this chapter, we note that the considerations of this and previous chapters are related to the mathematical theory of degeneration of systems of partial differential equations of a parabolic type (the simplest parabolic equation is one of one-dimensional (1D) heat conductivity in a uniform medium: $\partial u/\partial t = \partial^2 u/\partial x^2$). Indeed, the system of equations in the previous section,

$$\mathbf{Hu} + \mathrm{A}(\mathbf{w}, \mathbf{u}) = \mathbf{f},$$
$$\mathbf{Pw} + \mathrm{B}(\mathbf{u}, \mathbf{w}) = \mathbf{q}$$

may be considered as the parabolic system in which the parabolic operator of the first subsystem of the equations is degenerated into the hyperbolic operator **Hu** because of the absence of terms with second-order derivatives of the unknown function **u** in this subsystem.

In the parabolic equation considered in this chapter, the term with a second-order derivative of unknown function exists, but it is "disappearing" because the coefficient ε before this term tends to zero: $\varepsilon \geq 0$, i.e., the one parameter family of the parabolic equations is considered with a vanishing parameter before the second-order derivative of the solution (Bianchini and Bressan 2005).

In other words, in Chap. 10, the parabolic degeneration was a given static property of the system of PDEs under consideration, whereas the parabolic degeneration in this chapter is a nonstationary property increasing along with a decrease in the viscosity coefficient ε.

As a matter of fact, for both types of parabolic degeneration, a classical (in the theory of PDEs) investigation of the solution on the basis of the strict positive definiteness of the quadratic form determined by the second-order terms of the equations under consideration is impossible.

A nonlinear equation with a small parameter before the second-order spatial derivative is being considered, i.e., a equation with a vanishing viscosity coefficient. This equation differs from the model equation of turbulence of Burgers because of the existence in the case of degeneration (i.e., in the case of a zero coefficient of viscosity) of solutions that periodically depend only on the linear combination of the coordinate and time and obey the *conditions of the type of conservation laws and dissipation at the discontinuities* (Landau and Lifshitz 1987) of the degenerated solution. Because of the consideration of self-similar solutions, we are to apply not only the theory of PDF (as was the case up to this chapter) but also the theory of ordinary differential equations (ODEs), e.g., Calogero (2008), Dinjun and Libang (2013), Simon (2012), and Sternberg (2013).

11.2 Considered Solutions and the Method of Approximating Them by Vanishing Viscosity

The classical solution $U_\varepsilon(t, y)$ of the form

$$U_\varepsilon(t, y) = u_\varepsilon(x), x \equiv y - ct$$

of the equation

$$\partial U_\varepsilon/\partial t + 1/2\partial\left(U_\varepsilon{}^2\right)/\partial y = U_\varepsilon - c + \varepsilon\,\partial^2 U_\varepsilon/\partial y^2, \varepsilon, c > 0 \tag{11.2}$$

will be called an *rw-solution* if

$$\sup u_\varepsilon(x) < \infty$$

$$-\infty < \mathrm{x} < \infty.$$

The function $u_\varepsilon(x)$ is the *profile* of $U_\varepsilon(t, y)$, $\varepsilon \geq 0$. Below, $d = 1$ (see previous section) because the same consideration is valid for every $d > 0$; $c > 0$ is fixed to be arbitrary.

The function $U_0(t, y)$ of the form

$$U_0(t,y) = u_0(x), x \equiv y - ct,$$

with the *periodic* in regard to x profile $u_0(x)$ by the finite number of breakpoints x_i on the period and finite limit values $u_0(x_i - 0)$, $u_0(x_i + 0)$ for every x_i, will be called a *shock rw-solution* of the equation

$$\partial U_0/\partial t + 1/2\partial\left(U_0{}^2\right)/\partial y = U_0 - c \tag{11.3}$$

if $U_0(t, y)$ is continuously differentiable by $x \neq x_i$ and obeys this equation, whereas

$$u_0(x_i - 0) + u_0(x_i + 0) = 2c, u_0(x_i - 0) > u_0(x_i + 0) \text{ for every } x_i. \tag{11.4}$$

Relations similar to this equality in (11.4) are known in hydrodynamics and in the theory of discontinuous solutions of nonlinear PDEs as the Rankine–Hugoniot conditions at a shock (or jump); the inequality is an analog of the dissipative condition of compression by going through a shock (Colombeau and Le Roux 1988; Golitsyn et al. 2014; Henkin and Shananin 2014; Landau and Lifshitz 1987; LeVeque 2002; Ladyjenskaya 1969; Rossinger 1987; Wesseling 2001).

The set of the rw-solutions of Eq. (11.2) for the given $\varepsilon > 0$ is denoted by M_ε. The set of the rw-solutions of Eq. (11.3) is denoted by M_0.

We consider *approximation of an arbitrary rw-solution* $U_0(t, y) \in M_0$ of Eq. (11.3) with a zero coefficient of viscosity $\varepsilon = 0$ by *rw-solutions* $U_\varepsilon(t, y) \in M_\varepsilon$ of Eq. (11.2) with $\varepsilon > 0$ by $\varepsilon \to 0$.

Because of the theorems about vanishing viscosity (Oleinik 1963), $U_\varepsilon(t, y) \to U_0(t, y)$ by the condition that the initial values of all viscous solutions (i.e., by $\varepsilon > 0$) are equal to the initial value of the nonviscous solution ($\varepsilon = 0$), which must be approximated by $\varepsilon \to 0$:

$$U_\varepsilon(t,y)|_{t=0} = U_0(t,y)|_{t=0}, \; -\infty < y < \infty, \varepsilon \geq 0$$

But the approximating viscous solutions, by this condition, are not determined by the argument $x \equiv y - ct$, i.e., they are not, generally speaking, of the same spatial–temporal structure as the approximated nonviscous solution, though this similarity of structures would be natural from the physical point of view.

To obtain the desirable approximation of a nonviscous rw-solution (i.e., shock wave) by the smooth viscous rw-solution, we need to find another way to choose the approximating solutions.

The arbitrary *rw-solution* $U_0(t, y) \in M_0$ will be approximated with $U_\varepsilon(t, y) \in M_\varepsilon$: $U_\varepsilon(t, y) \to U_0(t, y)$, $\varepsilon \to 0$, by the condition

$$\Phi(U_\varepsilon(t,y)) = \Phi\,(U_0(t,y)), \varepsilon > 0 \tag{11.5}$$

Here Φ is a *functional with the following properties*:

1^0—Φ is determined on every M_ε, $\varepsilon \geq 0$;

2^0—For every *rw-solution* $U_0(t, y) \in M_0$, the number $\varepsilon(U_0, \Phi)$ exists such that if $0 < \varepsilon < \varepsilon(U_0, \Phi)$ then there is the unique *rw-solution* $U_\varepsilon(t, y) \in M_\varepsilon$ satisfying (11.5);

3^0—For the sequence $U_\varepsilon(t, y)$ determined by condition (11.5),

$$\|U_0(t,y) - U_\varepsilon(t,y)\|_{Lp(\Omega)} \to 0, \text{by } \varepsilon \to 0, p \geq 1. \tag{11.6}$$

Ω is an arbitrary domain on the plane *yt*, mes $\Omega < \infty$; $Lp(\Omega)$ denotes the Banach space of functions summarized on Ω along with the degree p (Brezis 2010; Saks 2017).

The class of functionals with properties 1^0, 2^0, 3^0 is not empty; examples will be demonstrated below.

We obtain from Eq. (11.2) for every fixed $\varepsilon > 0$:

$$M_\varepsilon = \{U_\varepsilon(t,y,C)|U_\varepsilon(t,y,C) = u_\varepsilon\,(x,C), 1 \leq C < \infty\}, \text{where } x \equiv y - ct.$$

Here u_ε (x, C) is determined by the one-parametric family, with the parameter C, of the limited solutions

$$(u_\varepsilon\,(x,C), v_\varepsilon\,(x,C))$$

of the autonomous (synonym: dynamic) system of equations

$$\varepsilon\, dv/dx = (1 - v)(c - u) \tag{11.7}$$
$$du/dx = v.$$

The system (11.7) follows from the PDE (11.2) because the profile u_ε (x, C) obeys the equation

$$\partial u(y-ct)/\partial t + {}^{1}\!/\!_{2}\partial\left(u^2(y-ct)\right)/\partial y = u(y-ct) - c + \varepsilon\,\partial^2 u(y-ct)/\partial y^2,$$

which may be written in the form $(-c)\ du/dx + u\ du/dx = u - c + \varepsilon\ d^2u/dx^2$, or (11.7).

After multiplication by v, the first equation of system (11.7) can be rewritten as

$$\varepsilon v(1-v)^{-1}\ dv/dx = (c-u)du/dx$$

We shall transform the left and right part of this equation:

$$\begin{aligned}\varepsilon v(1-v)^{-1}\ dv/dx &= \varepsilon\ [(v-1)/(1-v) + 1/(1-v)]\ dv/dx\\ &= -\varepsilon\ dv/dx + \varepsilon(1-v)^{-1}\ dv/dx\\ &= \varepsilon\ d(-v - ln(1-v))/dv\end{aligned}$$

Then $d(v + \ln(1-v))/dx = 0.5\ \varepsilon^{-1}\ d(c-u)^2/dx$.

So, we have the first integral of system (11.7):

$0.5\ \varepsilon^{-1}(c-u)^2 - (v + \ln(1-v) + C) = 0$, $C > 0$ is the constant of integration.

It is seen from here that the limited solutions of the system (11.7) are represented on the phase plane uv by cycles [we denote them as $L(\varepsilon, C)$] determined by the equation

$$(2\ \varepsilon)^{-1}(u-c)^2 - (f\ (v) + C) = 0;$$

here

$$f(v) \equiv \ln(1-v) + v, 0 < C < \infty. \qquad (11.8)$$

These cycles are symmetric in regard to the straight line $u = c$ and fill the half-plane $v < 1$. The half-plane $v > 1$ is filled by the unlimited trajectories of (11.7) and these trajectories, in view of the given definition of the rw-solutions, are not considered by us.

From the system of Eq. (11.7) by $\varepsilon = 0$, we obtain that there are u_0 in M_0 with an arbitrary number of breakpoints on the period of the profile. But according to the theorem of Byhovsky (1966), if a periodic profile u_0 is the limit of the profiles u_ε by $\varepsilon \to 0$, then the distance between neighboring breakpoints of u_0 is equal to the period of u_0. For a similar u_0, the limit values on the left and on the right a breakpoint are the same for all breakpoints.

In view of (11.4), M_0 is the following family of functions $U_0(t, x, u^+)$ with the parameter u^+,

$$u^+ \equiv u_0(x_i + 0),\ -\infty < u^+ < c:$$

$$M_0 \equiv \{U_0(t,y,u^+) | U_0(t,y,u^+) = u_0(x,u^+), u_0(x,u^+) = u^+ + x \text{ for } 0 \le x \le u^- - u^+,$$
$$u_0(x + u^- - u^+, u^+) = u_0(x,u^+) \text{ by } x \equiv y - ct, \ -\infty < x < \infty\}; \quad (11.9)$$

here $u^- \equiv u_0(x_i - 0)$, $u - \ = 2c - u^+$; see (11.4).

The above relation $u_0(x, u^+) = u^+ + x$ follows from Eq. (11.7) by $\varepsilon = 0$, because $v = 1$ in this case.

The dependence of functions from M_ε, $\varepsilon \ge 0$, and their profiles and derivatives, on C or u^+ will be omitted when it is clear from the text.

11.2.1 Convergence of the Smooth Rolling Wave Solutions to a Shock Wave with the Same Period

Let us introduce the functional T determined on every M_ε, $\varepsilon \ge 0$, such that the value of T on the rw-solution $U_\varepsilon(t, y, C)$ is the period of its profile.

It follows from (11.9) that

$$T\left(U_0(t,y,u^+)\right) \downarrow 0 \text{ by } u^+ \to c, T\left(U_0(t,y,u^+)\right) \uparrow \infty \text{ by } u^+ \to \infty.$$

Because of (11.7) and (11.8) and the theorems about the limit under the integral sign, we have (Danford and Schwartz 1988)

$$T(U_\varepsilon(t,y,C)) = T^+(\varepsilon,C) + T^-(\varepsilon,C), T^+(\varepsilon,C) \equiv (2\varepsilon)^{1/2} \int_0^1 \phi(z, q_\varepsilon(C))dz$$

$$T^-(\varepsilon,C) \equiv (2\varepsilon)^{1/2} \int_0^1 \phi(z, -w_\varepsilon(C))dz, \phi(z,q) \equiv q(f(zq) + C)^{-1/2}(1 - zq)^{-1}.$$

Here $f(v) \equiv \ln(1 - v) + v$ (see 11.8); $T^+(\varepsilon, C)$ denotes the time interval of motion of the point (u, v), imaging the solution of the dynamic system (11.7) on the phase plane uv, along the upper part (where $v > 0$) of the cycle $L(\varepsilon, C)$; $T^-(\varepsilon, C)$ denotes the time interval of motion along the lower part (where $v < 0$) of $L(\varepsilon, C)$.

We have by $\varepsilon = \text{const}$ and $C \to \infty$

$$0 < q(\varepsilon,C) \equiv \max_{(u,v)\in L(\varepsilon,C)} v \uparrow 1 - 0, \ \ 0 < w(\varepsilon,C) \equiv -\min_{(u,v)\in L(\varepsilon,C)} v \uparrow \infty,$$
$$T(U_\varepsilon(t,y,C)) \uparrow \infty.$$

Here $q(\varepsilon, C)$ is the maximal value of the vertical coordinate v of the point (u, v) on the cycle $L(\varepsilon, C)$ of the phase portrait of system (11.7) on the phase plane uv; $-w(\varepsilon, C)$ is the minimal value of the vertical coordinate v of the point (u, v) on the cycle $L(\varepsilon, C)$.

We have

$$q(\varepsilon, C) \downarrow 0+, T(U_\varepsilon(t, y, C)) \downarrow 2\varepsilon^{1/2}\pi \quad \text{by} \quad \varepsilon = \text{const} \quad \text{and} \quad C \to +0 \tag{11.10}$$

$$f(q(\varepsilon, C)) = f(-w(\varepsilon, C)) = -C, \partial q(\varepsilon, C)/\partial C > 0 \text{ by } 0 < C < \infty, \varepsilon > 0.$$

To check that the functional $T(U_\varepsilon(t, y, C))$ allows us to determine the only solution of system (11.7) with the help of condition (11.5) by $\Phi = T$, we write (11.7) in polar coordinates and apply the theorem from Chouikha (2005) and Vorobyev (1962) about periods in the case of a center. According to this theorem the functional T increases strictly monotonically in C, $0 < C < \infty$. It takes all values between $2\,\varepsilon^{1/2}\pi$ and ∞. So, Eq. (11.5), with T as the functional for Φ, has the only solution $U_\varepsilon(t, y) \in M_\varepsilon$ by arbitrary fixed $U_0(t, y) \in M_0$ and ε, $0 < \varepsilon < \varepsilon(U_0, T)$.

Also, the functional T has the property 3^0 accompanying condition (11.5) by $\Phi = T$: namely the rw-solutions $U_\varepsilon(t, y) \in M_\varepsilon$ determined by this condition approximate $U_0(t, y) \in M_0$ by $\varepsilon \to 0$.

Indeed, let us fix any $U_0(t, y) \in M_0$. Let us denote by C_ε the value of parameter C of the single rw-solution $U_\varepsilon(t, y) \in M_\varepsilon$, which obeys condition (11.5) by $\Phi = T$ and fixed above rw-solution in the right part of the condition. The dependence of C_ε on u^+ is omitted here. It follows from (11.9) and (11.10) by $\varepsilon \to 0$:

$$\begin{gathered} C_\varepsilon \to \infty, q(\varepsilon, C_\varepsilon) \to 1, w(\varepsilon, C_\varepsilon) \to \infty, T^-(\varepsilon, C_\varepsilon) \to 0, \\ \min_{(u,v)\in L(\varepsilon,C)} u \to u^+, \max_{(u,v)\in L(\varepsilon,C)} u \to u^-. \end{gathered} \tag{11.11}$$

Now we take in account that function $u_\varepsilon(x, C) - u_0(x, u^+)$ decreases monotonically by $x \in (0, u^- - u^+)$ when ε, C, u^+ are fixed (let us remember that $u^- - u^+$ is the period of $u_0(x, u^+)$ and $du_\varepsilon(x, C)/dx = v_\varepsilon(x, C) < 1$, whereas $du_0(x, u^+)/dx = 1$). Thus, there is convergence $U_\varepsilon(t, y, C_\varepsilon) \to U_0(t, y, u^+)$ when $\varepsilon \to 0$ by condition (11.5), this convergence is uniform in regard to points of the plane ty, which do not belong to the sum of the arbitrary narrow bands near the lines of brakes of the nonviscous solution $U_0(t, y, u^+)$.

$$\|U_0(t, y, u^+) - U_\varepsilon(t, y, C_\varepsilon)\|\boldsymbol{C}(B_\delta) \to 0 \text{ by } \varepsilon \to 0 \tag{11.12}$$

where $\boldsymbol{C}(B_\delta)$ is the space of continuous functions on the domain B_δ

$$B_\delta = \{(t, y)| -\infty < t, y < \infty\} \backslash \sum \{(t, y)| kT_0 - \delta < y - ct < kT_0\}$$

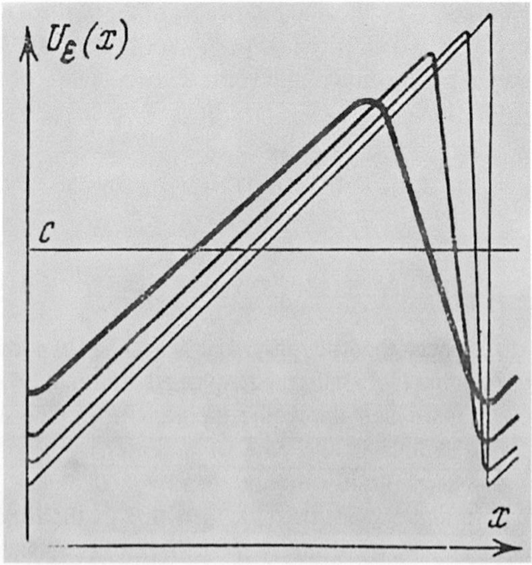

Fig. 11.1 Viscous rolling wave solutions (*rw-solutions*) approximate the shock rw-solution by vanishing viscosity and the condition of equality of periods

$$k \in (-\infty, +\infty)$$

$$T_0 = T(U_0(t, y, u^+)) = u^- - u^+, \; -\infty < u^+ < c, 0 < \delta < T_0.$$

Because $\delta > 0$ may be chosen arbitrary small, the convergence (11.6) follows from (11.11) and (11.12) (Fig. 11.1).

11.3 Vanishing Viscosity $\varepsilon \to 0$ by the Condition of Time of Motion $T^+(\varepsilon, C)$ of the Imaging Point with Velocity $V > 0$ Is Equal to the Period T_0 of the Shock Wave

The following notations and definitions will be used:

$u^{\min}(\varepsilon, C) \equiv \min u$ by $(u, v) \in L(\varepsilon, C)$; $u^{\max}(\varepsilon, C) \equiv \max u$ by $(u, v) \in L(\varepsilon, C)$;

$L(\varepsilon, C)$ denotes the cycle with the parameter C of the dynamic system (see Sect. 11.1)

$$\varepsilon \, dv/dx = (1 - v)\,(c - u),$$

$$du/dx = v;$$

the point *imaging* (i.e., *representing*) *a dynamic system* on the phase plane *uv* is point $(u(t), v(t))$ moving on this plane according to this system, i.e., $(u(t), v(t))$ is a solution of the system of its equations; $T^+(\varepsilon, C)$ is the time interval of motion of the imaging point (u, v) by $v > 0$ along the cycle $L(\varepsilon, C)$ with the parameter C:

$$T^{+}(\varepsilon, C) \equiv (2\varepsilon)^{1/2} \int_{0}^{1} \phi(z, q_{\varepsilon}(C))dz$$

$$\phi(z, q) \equiv q(f(zq) + C)^{-1/2}(1 - zq)^{-1}$$

$f(v) \equiv \ln(1 - v) + v$ (see 11.8).

T_0 is the period of the profile $u_0(x)$ of the nonviscous ($\varepsilon = 0$) solution, i.e., the shock wave, we are to approximate; $T_0 = u^- - u^+$; u^+ denotes the value of the shock wave's height after the jump by $x = x_j$: $u^+ \equiv u(x_j + 0)$, whereas $u^- \equiv u^-(x_j - 0)$, and $u^+ < u^-$ according to the dissipation condition on a jump (see (11.9)); $u^+ \equiv min\, u_0(x)$, $u^- \equiv max\, u_0(x)$ by $x \in [0, T_0]$.

The functional Φ for the choice of the viscous ($\varepsilon > 0$) solution approximating the nonviscous one (see Sect. 11.1) is

$$\Phi = T^{+}(\varepsilon, C).$$

We are to approximate here the nonviscous solution ($\varepsilon = 0$) with profile $u_0(x)$ and period T_0 by the following condition for the choice of the viscous solution ($\varepsilon > 0$):

$$T^{+}(\varepsilon, C) = T_0$$

Theorem *The extremes of the viscous wave obeying the condition $T^{+}(\varepsilon, C) = T_0$ approximate the corresponding extremes u^+, u^- of the shock rw-solution by $\varepsilon \to 0$.*

$$u^{\min}(\varepsilon, C_{\varepsilon}) \to u^{+}, u^{\max}(\varepsilon, C_{\varepsilon}) \to u^{-} \ by \ \varepsilon \to 0.$$

Proof To obtain the convergence of the viscous extremes $u^{\max}(\varepsilon, C_\varepsilon)$, $u^{\min}(\varepsilon, C_\varepsilon)$ to the nonviscous ones by vanishing viscosity and condition $T^{+}(\varepsilon, C) = T_0$, we are to clear out the formulas connecting extremes and $T^{+}(\varepsilon, C)$ with the integral and the phase portrait of the dynamical system under consideration. Because of this integral, namely

$$(2\,\varepsilon)^{-1}(u - c)^2 = f(v) + C, f(v) \equiv v + \ln(1 - v),$$

we have $u = c \pm (2\varepsilon(f(v) + C))^{1/2}$ by $f(v) + C \geq 0$, i.e., by $-w(C) < v < q(C)$, where $-w(C)$ is the negative root and $q(C)$ is the positive one of equation $f(v) + C = 0$. Function $f(v) < 0$ for $v \neq 0$ and $f(0) = 0$, therefore

$$u^{\max}(\varepsilon, C\varepsilon) = c + (2\varepsilon\, C\varepsilon)^{1/2}, \mathrm{u}^{\min}(\varepsilon, C\varepsilon) = c - (2\varepsilon\, C\varepsilon)^{1/2}.$$

The formula for $T^{+}(\varepsilon, C)$, needed here, may be obtained from the equation of the dynamical system written in the form

$$\varepsilon\,(dv/dx)\,(1-v)^{-1}(c-u)^{-1}=1.$$

Because of the symmetry of the phase portrait in regard to the vertical line $u = c$ at the phase plane uv, the time of motion—we denote this time by Tp—of the imaging point along every cycle by $v > 0$ may be expressed as Tp = 2Tpl where Tpl denotes the time of motion along the left part of the cycle by $v > 0$. At this part of a cycle, $c - u > 0$ for all of the cycles. Therefore

$$c-u=(2\varepsilon(f(v)+C))^{1/2},$$
$$\varepsilon\,(dv/dx)\,(1-v)^{-1}(2\varepsilon(f(v)+C))^{-1/2}=1.$$

After integration of both parts of the last equation from $x = 0$ up to $x = \mathrm{Tpl}$, one gets the following equation for Tpl:

$$\int_0^{\mathrm{Tpl}} v'(x)(1-\ v(x))^{-1}(f(v(x))+C)^{-1/2}dx=\mathrm{Tpl}(\varepsilon/2)^{-1/2}$$

where $v'(x) \equiv dv(x)/dx$.

To get a more explicit expression for Tpl, we go over to integration by the new variable q connected with x:

$$q=v(x)$$

where $v(x)$ is the component of solution ($u(x)$, $v(x)$) of the system of ordinary differential equations of the dynamical system (let us remember that the latter is replacing the original PDE by $x = y - ct$, $\varepsilon > 0$). Then

$$dq=v'(x)dx,\, v(0)=0,\, v(\mathrm{Tpl})=\mathit{max}\ v \text{ on the cycle } L(\varepsilon,C)\equiv q(\varepsilon,C)$$

because the maximum of the vertical coordinate v on $L(\varepsilon, C)$ takes place just after the time interval Tpl from the beginning of the motion along this cycle. As this maximum was denoted by $q(\varepsilon, C)$, we have the following expression for Tpl:

$$(\varepsilon/2)^{1/2}\int_0^{q(\varepsilon,C)}(1-q)^{-1}(f(q)+C)^{-1/2}dq=\mathrm{Tpl}.$$

We note that this integral has the feature by $q = q(\varepsilon, C)$ because $f(q(\varepsilon, C)) + C = 0$, and it is dependent on the parameter C not only directly under the sign of the root, but because of the upper limit of integration. (Let us remember that $q(\varepsilon, C)$ was determined as the positive solution of the equation $f(q)) + C = 0$). We need to consider the integral in more detail to investigate the vanishing viscosity.

Because
$(1-q)^{-1} \le (1-q(\varepsilon,C))^{-1}$, where $q(\varepsilon,C) \in (0,1)$,

the second multiplier only must be considered under the sign of the last integral. According to the rule of L'Hôpital,

$$\lim_{q\to q(\varepsilon,C)}(f(q)+C)/(q-q(\varepsilon,C)) = \lim_{q\to q(\varepsilon,C)} df(q)/dq$$
$$= \lim_{q\to q(\varepsilon,C)}(-(1-q)^{-1}+1) = -q(\varepsilon,C)/(1-q(\varepsilon,C)).$$

Therefore, the feature under the sign of the integral is $O((q-q(\varepsilon,C))^{-1/2})$, i.e., it is summarized with the degree $2-\delta$, $0<\delta\le 1$.

To apply this integral to study the nonviscous limit, we transform it into the form of the sum of two terms $I_1(C)$ and $I_2(C)$: $I_1(C)\to\infty$, $|I_2(C)| \le const$ by $C\to\infty$:

$$\text{Tpl} = (\varepsilon/2)^{1/2}\int_0^{q(\varepsilon,C)}((1-q)^{-1}-1)(f(q)+C)^{-1/2}dq$$
$$+(\varepsilon/2)^{1/2}\int_0^{q(\varepsilon,C)}(f(q)+C)^{-1/2}dq \equiv (\varepsilon/2)^{1/2}(I_1(C)+I_2(C)).$$

Here $f(q)\equiv q+\ln(1-q)$, $df(q)/dq = 1-(1-q)^{-1} = -((1-q)^{-1}-1)$,

$$d(f(q)+C)^{1/2}/dq = -0.5(f(q)+C)^{-1/2}\left((1-q)^{-1}-1\right),$$

$$I_1(C) \equiv \int_0^{q(\varepsilon,C)}\left((1-q)^{-1}-1\right)(f(q)+C)^{-1/2}dq = (-2)(f(q)+C)^{1/2}\Big|_{q=0}^{q=q(\varepsilon,C)}$$
$$= 2C^{1/2}$$

$$I_2(C) \equiv \int_0^{q(\varepsilon,C)}(f(x)+C)^{-1/2}dx.$$

So

$$\text{Tpl} = (\varepsilon/2)^{1/2}\left(2C^{1/2}+I_2(C)\right). \quad (11.13)$$

The integral I_2 has the particularities mentioned above, and we have to consider it to investigate vanishing viscosity. We go over to integration with the variable

$$x' = x/q(\varepsilon,C)$$
$$I_2(C) \equiv q(\varepsilon,C)\int_0^1 (f(q(\varepsilon,C)x)+C)^{-1/2}dx.$$

The previous notation for the variable of integration is used here.

To clear the behavior of $I_2(C)$ by $C\to\infty$ it is useful to consider

$$d(f(q(\varepsilon,C)x)+C)/d\,C = d/dC(\ln(1-q(\varepsilon,C)\,x)+q(\varepsilon,C)\,x+C)$$
$$= \frac{-q_C x}{1-q(\varepsilon,C)x}+q_C x+1$$
$$= \frac{-q(\varepsilon,C)q_C x^2+1-q(\varepsilon,C)x}{1-q(\varepsilon,C)x}, \quad q_C \equiv dq(\varepsilon,C)/dC.$$

We are going to exclude q_C from this expression of $d(f(q(\varepsilon, C)x) + C)/dC$. To do so, let us remember that $q = q(\varepsilon, C)$, $0 < q(\varepsilon, C) < 1$, is the only positive solution of the equation

$\ln(1 - q) + q + C = 0$, i.e., $\ln(1 - q(\varepsilon, C)) + q(\varepsilon, C) + C = 0$ for every $C > 0$. After differentiation we have $-(1 - q(\varepsilon, C))^{-1} q_C + q_C + 1 = 0$, $(-(1 - q(\varepsilon, C))^{-1} + 1)$ $q_C = -1$,

$$q_C = (1-q(\varepsilon,C))/q(\varepsilon,C).$$

Therefore,

$$d(f(q(\varepsilon,C)x)+C)/dC = (-(1-q(\varepsilon,C))x^2 - q(\varepsilon,C)x+1)/(1-q(\varepsilon,C)x)$$
$$= P_2(x)/(1-q(\varepsilon,C)x),$$
$$\text{where } P_2(x) \equiv -(1-q(\varepsilon,C))\,x^2 - q(\varepsilon,C)x+1.$$

As $dP_2(x)/dx = -2(1 - q(\varepsilon, C))\,x - q(\varepsilon, C) < 0$ by $x \in [0, 1]$, then $P_2(x)$ decreases on this interval from $P_2(0) = 1$ to $0 = P_2(1)$, i.e.,

$$sign P_2(x) = sign\, d\,(f(q(\varepsilon,C)x)+C)/dC > 0 \text{ by } x \in (0,1), q(\varepsilon,C) \in (0,1)$$

Therefore, by these values of x and $q(\varepsilon, C)$: $f(q(\varepsilon, C_2)x) + C_2 > f(q(\varepsilon, C_1)x) + C_1$ by $C_2 > C_1$, and the function under the sign of integral I_2 (C) decreases when C increases. Thus

$$|I_2(C)| \leq const \text{ by } C \to \infty. \tag{11.14}$$

Let us return to the condition of the choice of the viscous solution ($\varepsilon > 0$):

$$T^+(\varepsilon, C_\varepsilon) = T_0 = u^- - u^+$$

According to the above formula for Tpl after the choice $C = C_\varepsilon$ by this condition, $2\text{Tpl} = 2(\varepsilon/2)^{1/2}\,(2C^{1/2} + I_2(C)) = u^- - u^+$, or $(2\varepsilon C)^{1/2} + (\varepsilon/2)^{1/2} I_2(C) = u^-/2 - u^+/2$, $(2\varepsilon C_\varepsilon)^{1/2} + c = u^-/2 - u^+/2 + c - (\varepsilon/2)^{1/2} I_2(C_\varepsilon)$.

The left part of this equality may be replaced by $u^{max}(\varepsilon, C_\varepsilon) = c + (2\varepsilon\, C_\varepsilon)^{1/2}$, in the right part, $u^- + u^+ = 2c$ due to the condition on a jump and $u^+ = 2c - u^-$. So, due to the estimation $I_2(C) < const$ by $C \to \infty$ (see above)

$$u^{max}(\varepsilon, C_\varepsilon) = u^- + O\left(\varepsilon^{1/2}\right) \text{ by } \varepsilon \to 0.$$

The convergence $u^{min}(\varepsilon, C_\varepsilon) = c - (2\varepsilon\, C_\varepsilon)^{1/2} \to u^+$ by $\varepsilon \to 0$ and $T^+(\varepsilon, C_\varepsilon) = T_0 = u^- - u^+$ is demonstrated in the same way. ∎

Remarks (Uniform, in regard to $x \in [0, T_0]$, convergence $u_\varepsilon(x, C) \to u_0(x, u^+)$, $\varepsilon \to 0$.)

Because $du_\varepsilon(x, C)/dx = v_\varepsilon(x, C) < 1$ whereas $du_0(x, u^+)/dx = 1$ (Sect. 11.2), and therefore function $u_0(x, u^+) - u_\varepsilon(x, C)$ is increasing by $x \in [0, T_0]$, the maximal module of difference of the viscous $u_\varepsilon(x, C)$ and nonviscous $u_0(x, u^+)$ profiles is reached at the ends of the interval $[0, u^- - u^+]$. At these ends, the extremes of the considered solutions are located, and the difference between these extremes vanishes by $\varepsilon \to 0$ according to the above theorem

$$\sup_{x \in [0,T_0]} \left|u_0(x, u^+) - u_\varepsilon(x, C)\right| = max\left\{\left|u_0(0, u^+) - u_\varepsilon(0, C)\right|, \left|u_0(T_0, u^-) - u_\varepsilon(T_0, C)\right|\right\} \to 0 \text{ by } \varepsilon \to 0$$

11.4 Vanishing Viscosity by the Condition of Equality of Amplitudes $A_\varepsilon(C)$, $\varepsilon > 0$, of Viscous Solutions to the Amplitude A_0 of the Nonviscous Solution: $A_\varepsilon(C) = A_0$

We use the following notations for the maximal and minimal values of coordinate $u = u_\varepsilon(x, C)$ on the phase plane (u, v) of the trajectory of a cycle $L(\varepsilon, C)$:

$$u^{max}{}_\varepsilon(C) \equiv \max_{x \in [0,T\varepsilon]} u_\varepsilon(x, C), u^{min}{}_\varepsilon(C) \equiv \min_{x \in [0,T\varepsilon]} u_\varepsilon(x, C).$$

Here $u_\varepsilon(x, C)$ is the profile of the rw-solution of Eq. (11.1) and T_ε is the period of its profile. Its amplitude is denoted by

$$A_\varepsilon(C) \equiv u^{max}{}_\varepsilon(C) - u^{min}{}_\varepsilon(C)$$

The time of motion of the imaging point of the dynamic system (11.7) with the positive value of the velocity

$$du_\varepsilon(x, C)/dx = v_\varepsilon(x, C) > 0$$

is denoted by T_ε^+ [let us remember that x denotes the time in the case of system (11.7)]. Because the profile is increasing monotonically by $x \in [0, T_\varepsilon^+]$ and then, in the last part of the period, the profile is decreasing, its maximal value corresponds to $x = T_\varepsilon^+$.

Theorem *Let A_0denotes the amplitude of an arbitrary fixed shock rw-solution of Eq.* (11.3) *with zero viscosity $\varepsilon = 0$; let its profile be denoted by $u_0(x)$ and period by T_0.*

For every $\varepsilon > 0$, there is the only viscous rw-solution of Eq. (11.2) *such that:*

1. $A_\varepsilon(C_\varepsilon) = A_0$, $\varepsilon > 0$.
2. $\|u_\varepsilon(x, C_\varepsilon) - u_0(x)\|_{C[0,\, T_0]} \to 0$ *by* $\varepsilon \to 0$.

[The amplitude of this solution is denoted by $A_\varepsilon(C_\varepsilon)$ and the profile by $u_\varepsilon(x, C_\varepsilon)$, where $C = C_\varepsilon$is the only value of the parameter C of the cycle $(u_\varepsilon(x, C_\varepsilon), v_\varepsilon(x, C_\varepsilon))$ on the phase plain (u, v) of system (11.7) *by the condition* 1].

Proof From the integral of the dynamical system for the rw-solutions,
$(2\,\varepsilon)^{-1}(u - c)^2 = f(v) + C, f(v) \equiv \ln(1 - v) + v, 0 < C < \infty$ one can get

$$u^{max}{}_\varepsilon(C) = c + (2\varepsilon C)^{1/2}, u^{min}{}_\varepsilon(C) = c - (2\varepsilon C)^{1/2}$$

because $f(v) < 0$ for every v, save $v = 0, f(0) = 0$.
Therefore,

$$A_\varepsilon(C) \equiv u^{max}{}_\varepsilon(C) - u^{min}{}_\varepsilon(C) = 2(2\varepsilon C)^{1/2}.$$

It is clear now that the condition

$$A_\varepsilon(C) = A_0, \varepsilon > 0$$

(where A_0 is the amplitude of an arbitrary fixed nonviscous rw-solution) determines the only constant $C = C_\varepsilon$ and viscous rw-solution (its profile is denoted by $u_\varepsilon\,(x, C_\varepsilon)$) obeying this condition:

$$2(2\varepsilon C_\varepsilon)^{1/2} = A_0, C_\varepsilon = (1/8)\,\varepsilon^{-1}A_0{}^2$$

Let us note that because of the condition of equality of the viscous amplitudes to the nonviscous one, the viscous ($\varepsilon > 0$) extremes happen to be equal to the nonviscous extreme. Indeed, let us consider

$$u^{max}{}_\varepsilon(C_\varepsilon) - \mathrm{u}^- = c + (2\varepsilon C_\varepsilon)^{1/2} - u^- = (2\varepsilon C_\varepsilon)^{1/2} - (\mathrm{u}^- - \mathrm{c}) = ½A_\varepsilon(C_\varepsilon) - ½A_0 = 0.$$

Here we have replaced $(u^- - c)$ with $½A_0$ because of conditions (11.4) on a jump of a nonviscous rw-solution, namely $u_0(x_i - 0) + u_0(x_i + 0) = 2c$, i.e., $u^- + u^+ = 2c$.

The replacement is true because $u^- - c = c - u^+$, $A_0 = u^- - u^+$, $A_0 - u^- = -u^+$, $u^- - c = c + A_0 - u^-$, $2(u^- - c) = A_0$, $(u^- - c) = ½A_0$.

Analogically:

$$u^{\min}{}_{\varepsilon}(C_\varepsilon) - u^+ = (c - u^+) - (2\varepsilon C_\varepsilon)^{1/2} = u^- - c - (2\varepsilon C_\varepsilon)^{1/2} = ½A_0 - ½A_\varepsilon = 0.$$

Therefore,

$$u^{\max}{}_{\varepsilon}(C_\varepsilon) = u^-, u^{\min}{}_{\varepsilon}(C_\varepsilon) = u^+. \tag{11.15}$$

Thus, the functional of amplitude obeys the properties 1^0 and 2^0 of Sect. 11.1, and we are to demonstrate 3^0.

To do so, we apply (11.13) and (11.14) from Sect. 11.3 for the time of motion:

$$\text{Tpl} \equiv ½T_\varepsilon{}^+$$

of the imaging point of the dynamical system (11.7) along the left upper ($u < c, v > 0$) part of the cycle $L(\varepsilon, C)$:

$$\text{Tpl} = (\varepsilon/2)^{1/2}\left(2C^{1/2} + I_2(C)\right), I_2(C) \equiv \int_0^{q(\varepsilon,C)} (f(x) + C)^{-1/2} dx$$

where

$|I_2(C)| \le \text{const by } C \to \infty$.

Because of the symmetry of the phase portrait of system (11.7) and the above formulas

$$A_\varepsilon(C_\varepsilon) = A_0 = u^- - u^+ = T_0 = 2(2\varepsilon C)^{1/2}$$

we have (let us remember that index 0 denotes the amplitude and period of the nonviscous rw-solution)

$$T_\varepsilon{}^+ = (2\varepsilon)^{1/2}\left(2(\mathrm{C}_\varepsilon)^{1/2} + I_2(\mathrm{C})\right) = A_\varepsilon(C_\varepsilon) + (2\varepsilon)^{1/2} I_2(C_\varepsilon)$$

$$T_\varepsilon{}^+ - A_\varepsilon(C_\varepsilon) = (2\varepsilon)^{1/2} I_2(C_\varepsilon), T_\varepsilon{}^+ - A_0 = T_\varepsilon{}^+ - T_0$$

$$T_\varepsilon{}^+ - T_0 = (2\varepsilon)^{1/2} I_2(C_\varepsilon), |I_2(C)| \le \text{const by } C \to \infty \tag{11.16}$$

Let us demonstrate

$$||u_\varepsilon(x, C_\varepsilon) - u_0(x)||_{C[0,To]} = | u_\varepsilon(T_0, C_\varepsilon) - u_0(T_0) | \tag{11.17}$$

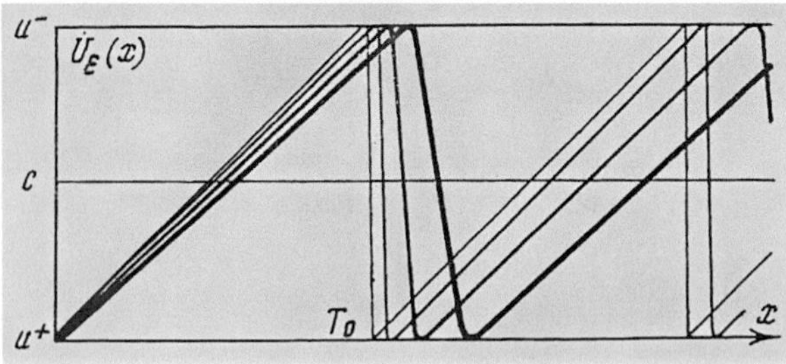

Fig. 11.2 Viscous rolling wave solutions (*rw-solutions*) approximate the shock rw-solution by vanishing viscosity and the condition of equality of amplitudes

Indeed, the function $u_0(x) - u_\varepsilon(x, C_\varepsilon)$ is increasing by $x \in [0, T_0]$ because $d(u_0(x) - u_\varepsilon(x, C_\varepsilon))/dx = 1 - v_\varepsilon(x, C_\varepsilon) > 0$ due to $v < 1$ everywhere in the domain occupied by cycles on the phase plain (Fig. 11.2).

Therefore,

$$\max_{x \in [0,T_0]} |\, u_\varepsilon(x, C_\varepsilon) - u_0(x) \,| \\ = \max\{|u_\varepsilon(T_0, C_\varepsilon) - u_0(T_0)|, |u_\varepsilon(0, C_\varepsilon) - u_0(0)|\}.$$

But $|u_\varepsilon(0, C_\varepsilon) - u_0(0)| = |u^{\min}{}_\varepsilon(C_\varepsilon) - u^+| = 0$, according to the second equality of (11.15), and we obtain (11.17). The first equality $u^{\max}{}_\varepsilon(C_\varepsilon) = u^-$ of (11.15) may be rewritten as

$$u^{\max}{}_\varepsilon(C_\varepsilon) = u_0(T_0)$$

because $u_0(T_0) = u^-$. Indeed, according to (11.9), $u_0(x, u^+) = u^+ + x$ for $0 \le x \le u^- - u^+$ (let us remember that usually we write $u_0(x)$ instead of $u_0(x, u^+)$; see the end of Sect. 11.1). We know that $u^{\max}{}_\varepsilon(C_\varepsilon) = u_\varepsilon(T_\varepsilon^+, C_\varepsilon)$ because $du_\varepsilon(x, C_\varepsilon)/dx = v_\varepsilon(x, C_\varepsilon) > 0$ by $0 < x < T_\varepsilon^+$ and $du_\varepsilon(x, C_\varepsilon)/dx < 0$ on the last part of the period of a viscous rw-solution. This allows us to express the right part of (11.17) in terms of the viscous profile only and then to apply the Lagrange theorem about an intermediate point (this point was denoted here by $x_\alpha = T_0 + \alpha\,(T_\varepsilon^+ - T_0)$, $0 < \alpha < 1$) and (11.16):

$\|u_\varepsilon\,(x, C_\varepsilon) - u_0(x)\|_{C[0,\,T_0]} = |\, u_\varepsilon(T_0, C_\varepsilon) - u_\varepsilon(T_\varepsilon^+, C_\varepsilon) \,| = u_\varepsilon(T_\varepsilon^+, C_\varepsilon) - u_\varepsilon(T_0, C_\varepsilon) = v_\varepsilon(x_\alpha, C_\varepsilon)(T_\varepsilon^+ - T_0) = (2\varepsilon)^{1/2} I_2(C_\varepsilon) \to 0$ by $\varepsilon \to 0$. ∎

Remarks The viscous rw-solutions approximate an arbitrary shock rw-solution by vanishing viscosity $\varepsilon \to 0$ in the norms $L_p\,(D)$, $mesD < \infty$, with the velocity of convergence of $O(|\varepsilon \ln \varepsilon|^{1/p})$, $p > 1$, by condition of the fixed value of the functional of the period or the fixed value of the functional of the amplitude (Novik 1971). The same paper investigated the convergence by $\varepsilon \to 0$ of the trajectories of the dynamical system representing on the phase plain the considered PDF in the case of self-similar solutions. Here was shown the strong difference of the considered model discontinuous hydrodynamic oscillations from the discontinuous relaxation oscillations (Ginoux and Litellier 2012).

11.5 Conclusion

The vanishing viscosity $\varepsilon \rightarrow 0$ was considered in the class of the self-similar solutions (of the type $U_\varepsilon(t, y) = u_\varepsilon(x)$, $x = y - ct$, $c =$ const is a parameter) of the Burgers model equation of turbulence, modified by us with an additional term without differentiation (see the Introduction to this chapter). (Similar terms are used sometimes in hydrodynamic models, e.g., the Saint Venant model). A modification was applied because by $\varepsilon = 0$ in the Burgers equation, the nontrivial self-similar solutions are absent. The question arises: What is the way to approximate an arbitrary fixed discontinuous self-similar solution of the Burgers modified equation with the smooth solutions of this equation with $\varepsilon > 0$? Usually, by consideration of vanishing viscosity, the approximating viscous solutions have being chosen with the same initial conditions as the discontinuous (i.e., shock wave) solution with $\varepsilon = 0$, which must be approximated by smooth solutions ($\varepsilon > 0$) of the same spatial–temporal structure. But in the case of self-similarity, this approach is impossible because a sequence of self-similar functions of (t, y)—with the same initial conditions (i.e., initial values) for all of the functions—includes only one function! But the hydrodynamic regimes with self-similarity are of interest and importance from the physical and mathematical points of view, and we are to replace the approach with the same initial values, which happened to be too restrictive in the case of self-similarity, by an approach with the same value of some functional $\Phi(U_\varepsilon(t, y)) = \Phi(U_0(t, y))$, $\varepsilon > 0$, determined on $U_\varepsilon(t, y)$, $\varepsilon \geq 0$. In the sections of this chapter we have proved that the functional of the period of the profile $u_\varepsilon(x)$; the functionals of some characteristics of motion of the dynamic system representing the Burgers modified equation on the phase plane (u, v), $du/dx = v$, due to self-similarity (Sect. 11.3); and the functional of amplitude may be applied to approximate the arbitrary fixed discontinuous solution with smooth viscous solutions by vanishing viscosity.

References

Bec, J., & Khanin, K. (2007). Burgers turbulence. *Physics Reports, 447*((1–2), 1–66.

Bianchini, S., & Bressan, A. (2005). Vanishing viscosity solution of non-linear hyperbolic systems. *Annals of Mathematics, 161*, 223–342.

Brezis, H. (2010). *Functional analysis, Sobolev spaces and partial differential equations* (p. 599). Heidelberg: Springer. https://doi.org/10.1007/978-0-387-70914-7.

Byhovsky, E. B. (1966). *About the self-similar solutions of the type of a propagating wave of some quasi-linear differential equation and the system of equations describing the flow of water in a sloping channel.* (In Russian, Прикладная Математика и Механика, Prikladnaya Matematika i Mechanika, V. 30, Issue 2).

Calogero, F. (2008). *Isochronous systems*. Oxford: Oxford University Press.

Chouikha, A. R. (2005). Monotonicity of the period function for some planar differential systems. Part I: Conservative and quadratic systems. *Applicationes Mathematicae, 32*(3), 305–325.

Colombeau, J. F., & Le Roux, A. Y. (1988). Multiplication of distributions in elasticity and hydrodynamics. *Journal of Mathematical Physics, 29*((2), 315–319.

Danford, N., & Schwartz, J. (1988). *Linear operators, part I: General theory* (Vol. I, p. 872). Hoboken: Wiley.

Dinjun, L., & Libang, T. (2013). *Qualitative theory of dynamical systems, Advanced series in dynamical systems*. Singapore: World Scientific. https://doi.org/10.1142/1914. p. 272.

Dressler, R. F. (1949). Mathematical solution of the problem of roll-waves in inclined open channels. *Communications on Pure and Applied Mathematics, 2*(2–3), 149–194.

Ginoux, J.-M., & Litellier, C. (2012). Van der Pol and the history of relaxation oscillations: Toward the emergence of a concept. *Chaos: An Interdisciplinary Journal of Nonlinear Science, 22*(2), 023120.

Golitsyn, G. S. et al. (eds). (2014). *Turbulence, atmosphere and climate dynamics*. In Collected papers of the international conference dedicated to the memory of academician A.M. Obukhov (p. 696), Moscow, GEOS.

Henkin, G. M., & Shananin, A. A. (2014). Cauhy–Gelfand problem for quasilinear conservation law. *Bulletin des Sciences Mathématiques, 138*, 783–804.

Ladyjenskaya, O. A. (1969). *The mathematical theory of viscous incompressible flow*. New York/London: Gordon and Breach, Science Publications. (Translated from Russian, Moscow: Gos. Izdat. Fiz-Mat Lit. 1961).

Landau, L. D., & Lifshitz, E. M. (1987). *Fluid mechanics* (Vol. 6, 2nd ed.). Oxford: Butterworth–Heinemann.

LeVeque, R. (2002). *Finite volume methods for hyperbolic problems*. Cambridge: Cambridge University Press.

Novik, O. B. (1971). *Model description of rolling waves*. (In Russian: Прикладная математика и механика, Vol.35, Issue 6. *Journal of RAS*, Moscow. English Translation of the Journal: *Applied Mathematics and Mechanics*. Elsevier).

Oleinik, O. A. (1963). Discontinuous solutions of nonlinear differential equations. *American Mathematical Society Translations: Series, 2*(26), 95–172.

Rossinger, E. E. (1987). *Generalized solutions of nonlinear differential equations*. Amsterdam: Elsevier.

Saks, P. (2017). *Techniques of functional analysis for differential and integral equations* (p. 320). Amsterdam: Elsevier.

Simon, P. L. (2012). *Differential equations and dynamical systems*. Budapest: Department of Applied Analysis and Computational Mathematics, Institute of Mathematics, Eötvös Lorand University.

Sternberg, S. (2013). *Dynamical systems*. New York: Dover.

Vorobyev, A. P. (1962). About periods of solutions in the case of a center (In Russian). *Doklady Academy of Science of Belorusskoy SSR 6*(5).

Wesseling, P. (2001). *Principles of computational fluid dynamics*. Heidelberg: Springer.

Chapter 12
Conclusion

Abstract Geodynamic behaviour of our planet, i.e. earthquakes, seaquakes, tsunamis, geomagnetic storms etc., and their impacts on biosphere (including our brain and heart, see the previous Chapters), are the determining conditions of development of civilazation. In the concluding Chapter of the book about the mathematical geo-biophysics, the authors formulate the solved and mostly unsolved related problems.

Keywords Geodynamics · Main physical principles · Mathematical investigations · Coherence · Applications

In this book about geophysical fields and their impacts on humans, we have discussed both of these problems: our geophysical, biophysical, computational, and mathematical findings; and practical recommendations for land and marine (tsunami) seismological observations (see Chap. 3), as well as selection of operating personnel, e.g., with a relatively short restoration time of the brain's electromagnetic (EM) process coherence after it decreases during geomagnetic storms (see Chap. 6).

On the basis of the known geographical correlation between high-seismicity lithosphere zones and ones with low-resistivity structures, and applying the magneto-thermo-elasticity theory, we have formulated and mathematically investigated a model of magnetic location of a possible earthquake epicenter area. The model may not be of a well-investigated mathematical type, because of the different natures (EM, elastic) of the physical fields included and, therefore, the different types of the included differential operators; this just results in absence of a type in the model as a whole and the necessity of mathematical investigations (see Chap. 10). The calculations with this model demonstrate the observed seismo-EM phenomenon of magnetic location of a possible earthquake epicenter area (see Chap. 2). In the course of our seismo-EM research, we have demonstrated that a long EM wave inherits the spatial structure of the seismic wave during seismo-EM interaction by earthquake preparation. The effect of magnetic location of the possible epicenter area and the effect of seismic modulation of the EM wave are stable against uncertainty of the geometric and physical input data in the model. It would be of

O. Novik et al., *Electromagnetic Geophysical Fields*,
https://doi.org/10.1007/978-3-319-98461-2_12

interest to consider different geological and geophysical conditions of the formation of the magnetic location and to answer, theoretically and experimentally, the question about the possibility to "see" the magnetic location of the epicenter area of a possible earthquake from above the earth (through spectral "windows" in the ionosphere).

According to the main physical principles and geophysical data, we have formulated a mathematical model of the seismo-hydro-EM-temperature-geophysical field interaction (in the presence of gravity) arising in the lithosphere–ocean–atmosphere medium as a result of a weak seismic excitation (see Chap. 3) in the upper mantle under the ocean. The magnitudes of the computed seismic, EM, temperature, and hydrodynamic wave amplitudes, and the main frequencies and velocities, including the delay in regard to a submarine earthquake beginning, are of the orders of magnitude observed and published by other authors. This theory and the measurements by other authors were applied to determine a scheme of a multidisciplinary vertically distributed (from the seafloor up to the ionosphere) tsunami precursors detection system. On the basis of the field interaction theory and the computational methods described, it is desirable to consider different modifications of tsunami precursor–detection methods and devices, taking into account the different relief of the sea bottom and its hydrodynamic regime (turbulence). There are difficult problems with joint prognostic interpretation of data of different natures, such as submarine and satellite data (influenced by a hard sea, turbulence, ionosphere–magnetosphere plasma activity, etc.). There are questions of distortionless transmission of the data to regional and local prognostic centers and processing of the measured data, as well as the energy supply for sea bottom, ocean buoy, and balloon equipment, etc. (see Chap. 3). It is necessary to develop applications of methods of theoretical and experimental physics, and a mathematical theory for optimization of the engineering and economic problems involved in effective land and marine seismic prognosis.

Let us turn to the impact of the geophysical field dynamics on humans (see the title of this book on mathematical geo-biophysics).

We have investigated the influence of magnetic storms (MSs; also known as geomagnetic storms), namely the decrease in the coherence of the brain's EM processes caused by an MS. We needed to determine the part of the electroencephalographic (EEG) records with decreased coherence, for the sets of all of our records without the influence of an MS and for the sets with such an influence. To define the records with decreased coherence, we introduced a threshold parameter value; coherence with a smaller threshold size was determined as decreased. So we introduced a parameter of the storms algorithm to distinguish the records with decreased coherence. According to our experience of the EEG investigations, coherence values of 0.2 or 0.3 were effective to extract the records with a decreased value of the coherence function (in brief, *coherence*). The details are given in Chap. 6, including the possibility of testing with sequentially increasing values of the threshold, e.g., beginning from 0.1, and so on. It was shown that two values of the threshold parameter Th ($Th = 0.3$, $Th = 0.2$) of the algorithm for distinguishing the two sets of EEG records, i.e., the set without the influence of an MS and the set with such an influence, yielded consistent results. From these results, one can see that the coherence of the frontal polar and occipital theta rhythm brain bioelectrical oscillations in healthy persons at work on a computer is markedly decreased during (or even a day after) a moderate MS, whereas no deviations in the blood pressure, respiratory rate, or electrocardiographic (ECG) features are registered. It is desirable to determine the characteristics of the type of MS that has the greatest effect, including the decrease in coherence, as described. The quantitative electro-neurophysiological mechanism of a similar brain reaction to geomagnetic storms is unknown.

We turn now to the reaction of the heart. The results of standard cardiological measurements in a group of healthy young volunteers (students at the Third Medical Higher Education Institute in Moscow) in 2005 have been described. We showed that a number of the parameters that are widely applied in medicine for characterization of the functioning of the heart (see Chap. 8) experience distinct changes under the influence of a moderate geomagnetic storm, before returning to their background values within 2 days after the storm.

To undertake mathematical justifications of the above geophysical results, we have shown that systems of partial differential equations (PDEs) applied by us for description of the seismo-EM interaction during preparation of earthquakes, being without a type according to the classical typing of PDEs in mathematical physics, need to be investigated because similar models are used in different domains of physics when nonstationary interactions of different fields are considered (see Chap. 9). The general structure of these systems of PDEs—we call them string–diffusion systems of equations (SDSEs)—may be characterized as coupling of the hyperbolic and parabolic subsystems by the operators of interaction of physical fields of different natures (see Sect. 9.1).

Some results regarding the correct solvability of the initial boundary value problems [well-posed problems for the SDSEs (see Chap. 9)] were obtained, e.g., by grid approximation of the integro-differential ("memory" of the medium) system with hyperbolic and parabolic operators of the second order.

We proved (in Chap. 10) that for the Galerkin approximation of the initial boundary value problem of the nonlinear magneto-thermo-elasticity system, the solution exists in an arbitrary finite time interval $[0, T]$. Moreover, we proved that the functional energy norm of the Galerkin approximate solution—for every $0 \leq t \leq T$ and an arbitrary number of the elements of the Galerkin basis—is bounded with the value determined by the physical characteristics of the material, the Sobolev norms of the considered fields by $t = 0$, and the norms of the sources of the fields.

Earthquakes, seaquakes, and tsunamis are formed by fields of different natures, and some related mathematical results have been mentioned above. Another particularity of similar phenomena, accompanied by transfers of huge portions of energy, is discontinuous changes in the characteristics of the fields involved and difficulties of application of smooth functions for description. That is the reason for use of the smoothing method for vanishing viscosity. The vanishing viscosity $\varepsilon \to 0$ was considered in the class of self-similar solutions (of the type $U_\varepsilon(t,y) = u_\varepsilon(x)$, $x = y - ct$, $c = \text{const}$ as a parameter) of the Burgers model equation of turbulence, modified by us with an additional term without differentiation (see Introduction to in Chap. 11). The modification was applied because with $\varepsilon = 0$ in the Burgers equation, nontrivial self-similar solutions are absent. The question arises: What is the way to approximate an arbitrary fixed discontinuous self-similar solution of the Burgers modified equation by smooth solutions of this equation with $\varepsilon > 0$? Usually, with consideration of vanishing viscosity, approximating viscous solutions are chosen with the same initial conditions as a discontinuous (i.e., shock wave) solution with $\varepsilon = 0$, which should be approximated by smooth solutions ($\varepsilon > 0$) of the same spatial–temporal structure. But in the case of self-similarity, this approach is impossible because a sequence of the self-similar functions of (t,y)—with the same initial conditions (i.e., initial values) for all of the functions—includes only one function! But hydrodynamic regimes with self-similarity are of interest and importance from the physical and mathematical points of view, and an approach with the same initial values, which turned out to be too restrictive in the case of self-similarity, needs to be replaced by an approach with the same value of some functional $\Phi(U_\varepsilon(t, y)) = \Phi(U_0(t, y))$, $\varepsilon > 0$, determined on $U_\varepsilon(t,y)$, $\varepsilon \geq 0$. We proved (in Chap. 11) that the functional of the period of the profile $u_\varepsilon(x)$; the functionals of some characteristics of motion of the dynamic system representing the Burgers modified equation on the phase plane (u,v), $du/dx = v$, due to self-similarity (see Sect. 11.3); and the functional of amplitude may be applied to approximate the arbitrary fixed discontinuous solution with smooth viscous solutions for vanishing viscosity.

Here we have briefly described the main results of this work, and the problems that remain unsolved. Other directions for research to further develop the content of this book are described in the relevant chapters.

Index

O. Novik et al., *Electromagnetic Geophysical Fields*,
https://doi.org/10.1007/978-3-319-98461-2

T

MIX
Papier aus verantwortungsvollen Quellen
Paper from responsible sources
FSC® C105338

If you have any concerns about our products,
you can contact us on
ProductSafety@springernature.com

In case Publisher is established outside the EU,
the EU authorized representative is:
Springer Nature Customer Service Center GmbH
Europaplatz 3, 69115 Heidelberg, Germany

Printed by Libri Plureos GmbH
in Hamburg, Germany